Progress in Inflammation Research

Series Editor

Advisory Board

Forthcoming titles:

Chemokine Biology – Basic Research and Clinical Application, Volume 2: Pathophysiology of Chemokines, K. Neote, G. L. Letts, B. Moser (Editors), 2006
Toll-like Receptors in Inflammation, L. A. J. O'Neill, E. Brint (Editors), 2006
The Hereditary Basis of Rheumatic Diseases, R. Holmdahl (Editor), 2006
Endothelial Dysfunction and Inflammation, A. Graham (Editors), 2006
In vivo Models of Inflammation, 2nd edition, C. S. Stevenson, D. W. Morgan, L. A. Marshall (Editors), 2006

(Already published titles see last page.)

Chemokine Biology – Basic Research and Clinical Application
Volume I: Immunobiology of Chemokines

Bernhard Moser
Gordon L. Letts
Kuldeep Neote

Editors

Birkhäuser Verlag
Basel · Boston · Berlin

Editors

Bernhard Moser
Institute of Cell Biology
University of Bern
Baltzerstrasse 4
3012 Bern
Switzerland

Gordon L. Letts
NitroMed, Inc.
12 Oak Park Drive
Bedford, MA 01730
USA

Kuldeep Neote
Bone and Inflammation
Eli Lilly and Company
Lilly Corporate Center
Indianapolis, IN 46285
USA

A CIP catalogue record for this book is available from the Library of Congress, Washington D.C., USA

Bibliographic information published by Die Deutsche Bibliothek
Die Deutsche Bibliothek lists this publication in the Deutsche Nationalbibliografie;
detailed bibliographic data is available in the internet at http://dnb.ddb.de

ISBN-10: 3-7643-6825-X Birkhäuser Verlag, Basel – Boston – Berlin
ISBN-13: 978-3-7643-6825-8 Birkhäuser Verlag, Basel – Boston – Berlin

Part of Springer Science+Business Media
Printed on acid-free paper produced from chlorine-free pulp. TCF ∞
Cover design: Markus Etterich, Basel
Cover illustration: see page 159. With friendly permission of Osamu Yoshie.
Printed in Germany
ISBN-10: 3-7643-6825-X
ISBN-13: 978-3-7643-6825-8

e-ISBN: 3-7643-7423-3

9 8 7 6 5 4 3 2 1

www.birkhauser.ch

Contents

List of contributors

William W. Agace, Immunology Section, Lund University, BMC I-13, 22184 Lund, Sweden; e-mail: william.agace@med.lu.se

Marco Baggiolini, Università della Svizzera Italiana, Via Lambertenghi 10A, 6904 Lugano, Switzerland; e-mail: baggiolini@unisi.ch

Basil O. Gerber, Division of Allergology, Dept. of Rheumatology & Clin. Immunology/Allergology, Sahlihaus 1, Inselspital, 3010 Berne, Switzerland

Bernhard Homey, Department of Dermatology, Heinrich-Heine-University, Moorenstr. 5, 40225 Düsseldorf, Germany;
e-mail: bernhard.homey@uni-duesseldorf.de

Chenggang Jin, Division of Rheumatology, Department of Internal Medicine and the Interdisciplinary Group in Immunology, University of Iowa College of Medicine, EMRB 340F, Iowa City, IA 52242, USA

Paul Kubes, Immunology Research Group, Department of Physiology and Biophysics, University of Calgary, 3330 Hospital Drive N.W., Calgary, Alberta, T2N 4N1, Canada; e-mail: pkubes@ucalgary.ca

Antonio Lanzavecchia, Institute for Research in Biomedicine, Via Vincenzo Vela 6, 6500 Bellinzona, Switzerland; e-mail: lanzavecchia@irb.unisi.ch

Laura Lasagni, Center of Excellence for Research, Transfer and High Education DENOthe of the University of Florence, 50134 Italy; e-mail: l.lasagni@dfc.unifi.it

Lixin Liu, Immunology Research Group, Department of Physiology and Biophysics, University of Calgary, 3330 Hospital Drive N.W., Calgary, Alberta, T2N 4N1, Canada

Alexandra Lucas, BioTherapeutics Research Group/Vascular Biology Reasearch Group, Robarts Research Institute, and Dept. Microbiology and Immunology, University of Western Ontario, London, Ontario, Canada N6G 2K5;
e-mail: arl@robarts.ca

Charles R. Mackay, Garvan Institute of Medical Reasearch, Darlinghurst NSW 2010, Australia

Alfonso Martín-Fontecha, Institute for Research in Biomedicine, Via Vincenzo Vela 6, 6500 Bellinzona, Switzerland; e-mail: alfonso.martin-fontecha@irb.unisi.ch

Carlos Martínez-A., Department of Immunology and Oncology, Centro Nacional de Biotecnología/CSIC, Darwin, 3, Campus de Cantoblanco, 28049 Madrid, Spain

Grant McFadden, BioTherapeutics Research Group, Robarts Research Institute, and Dept. Microbiology and Immunology, University of Western Ontario, London, Ontario, Canada N6G 2K5; e-mail: mcfadden@robarts.ca

Dana McIvor, BioTherapeutics Research Group/Vascular Biology Reasearch Group, Robarts Research Institute, and Dept. Microbiology and Immunology, University of Western Ontario, London, Ontario, Canada N6G 2K5

Mario Mellado, Department of Immunology and Oncology, Centro Nacional de Biotecnología/CSIC, Darwin, 3, Campus de Cantoblanco, 28049 Madrid, Spain;
e-mail: mmellado@cnb.uam.es

Craig T. Morita, Division of Rheumatology, Department of Internal Medicine and the Interdisciplinary Group in Immunology, University of Iowa College of Medicine, EMRB 340F, Iowa City, IA 52242, USA; e-mail: craig-morita@uiowa.edu

Bernhard Moser, Institute of Cell Biology, University of Bern, Baltzerstrasse 4, 3012 Bern, Switzerland; e-mail: bernhard.moser@izb.unibe.ch

Joost J. Oppenheim, Laboratory of Molecular Immunoregulation, Intramural Research Support Program, Building 560, Room 21-89A, Frederick, MD 21702-1201, USA; e-mail: oppenhei@ncifcrf.gov

José Miguel Rodríguez-Frade, Department of Immunology and Oncology, Centro Nacional de Biotecnología/CSIC, Darwin, 3, Campus de Cantoblanco, 28049 Madrid, Spain; e-mail: jmrfrade@cnb.uam.es

Paola Romagnani, Center of Excellence for Research, Transfer and High Education DENOthe of the University of Florence, 50134 Italy;
e-mail: p.romagnani@dfc.unifi.it

Serigio Romagnani, Center of Excellence for Research, Transfer and High Education DENOthe of the University of Florence, 50134 Italy;
e-mail: s.romagnani@dmi.unifi.it

Federica Sallusto, Institute for Research in Biomedicine, Via Vincenzo Vela 6, 6500 Bellinzona, Switzerland; e-mail: federica.sallusto@irb.unisi.ch

Mariagrazia Uguccioni, Institute for Research in Biomedicine, 6500 Bellinzona, Switzerland

Osamu Yoshie, Department of Microbiology, Kinki University School of Medicine, Osaka-Sayama, Osaka 589-8511, Japan; e-mail: o.yoshie@med.kindai.ac.jp.

Ning Zhang, Laboratory of Molecular Immunoregulation, Intramural Research Support Program, Building 560, Room 21-89A, Frederick, MD 21702-1201, USA

Preface

The discovery of interleukin-8 close to 20 years ago initiated a new field of research touching on many aspects of immunology and inflammation. Interleukin-8 is just one member of a large class of structurally-related chemoattractant proteins, known as chemokines. Chemokines are involved in the traffic control of leukocytes, which bear the corresponding chemokine receptors on their surfaces. They are the largest family of cytokines in the human genome. The discovery of chemokines and chemokine receptors has been largely fueled by the human genome sequencing efforts. To date, there are more then 45 known chemokines and approximately 17 receptors.

Chemokine research over the last two decades has focused on their role in leukocyte migration. It is now clear that chemokines affect all aspects of immunology and contribute to the pathology of a large number of inflammatory and immune-mediated diseases, such as rheumatoid arthritis, pulmonary inflammatory diseases and multiple sclerosis. Their fundamental contributions to chronic inflammatory diseases make them a principal target for the development of novel, anti-inflammatory therapeutics. More recently, it has become apparent that chemokines have an essential role in diverse processes distinct from their function in immunity, including tumor cell growth and metastasis, atherosclerosis and angiogenesis. This book gives a state-of-the-art account of recent developments in this field in the form of summaries written by highly regarded experts.

Volume I is focused on basic principles and progress in chemokine biology. The emphasis is on the role of chemokines in leukocytes function and on their role in dendritic cell biology. In addition, chemokine receptor signaling and natural antagonism of the receptors is covered. Finally aspects of chemokine biology, as pertains to endothelial cells and angiogenesis, are discussed.

Volume II deals with issues related to the pathophysiology of chemokines, chemokine-related drug development and potential therapeutic applications. It is also published in the book series *Progress in Inflammation Research* and is entitled *Chemokine Biology – Basic Research and Clinical Application. Volume II: Pathophysiology of Chemokines* (2006, Birkhäuser, ISBN 3-7643-7195-1). These books provide both introductory and novel information for a broad readership, including clinicians and biomedical scientists.

September 2005

Bernhard Moser
Gordon L. Letts
Kuldeep Neote

Introduction

Introduction

Marco Baggiolini

Università della Svizzera Italiana, Via Lambertenghi 10A, 6904 Lugano, Switzerland

The beginning

In the first week of December 1987, two papers (one from the old, the other from the new world) presented the partial sequence of a novel protein, which was isolated from the culture supernatants of stimulated human monocytes and acted on neutrophil leukocytes. It was originally called NAF (neutrophil activating factor) [1], or MDNCF (monocyte-derived neutrophil chemotactic factor) [2]. At about the same time, two other laboratories reported the isolation of what turned out to be the same protein [3, 4]. The name was changed to NAP-1 (neutrophil-activating peptide number one) in the wise expectation to find analogues, but the new chemo-attractant became widely known by the fashionable and rather inappropriate name of interleukin-8 (IL-8).

After establishing the sequence, we rushed to a full analysis of the biological properties of IL-8 and found that its pattern of activity was qualitatively identical to that of known chemo-attractants for leukocytes, like the complement fragment C5a and N-formylmethionyl peptides [5]. The only difference was that IL-8 was selective for neutrophils, whereas the other attractants were non-specific. The effects of IL-8 were prevented by pretreatment of the cells with *Bordetella pertussis* toxin, a clear indication that they were mediated by a G-protein coupled receptor [5]. The initial observations, which were summarised in a JCI "Perspective" [6], attracted much interest. We needed large quantities of pure IL-8, which was produced biologically [7] and by chemical synthesis [8], and we concentrated on the study of IL-8 structure–activity relationships and, together with many others laboratories, on the search for IL-8-related chemokines.

In a decade of mining, human chemokines surfaced as a mega-family of 50 or so ligands and 20 receptors, all involved in leukocyte traffic [9]. The chemokines rapidly became a hot issue in immunology, pathology and medicine. Their biological relevance is perhaps best emphasised by the multiple interactions of viruses with the chemokine system, which evolved the expression of chemokines, receptors, antago-

Chemokine Biology – Basic Research and Clinical Application, Volume I
edited by Bernhard Moser, Gordon L. Letts and Kuldeep Neote

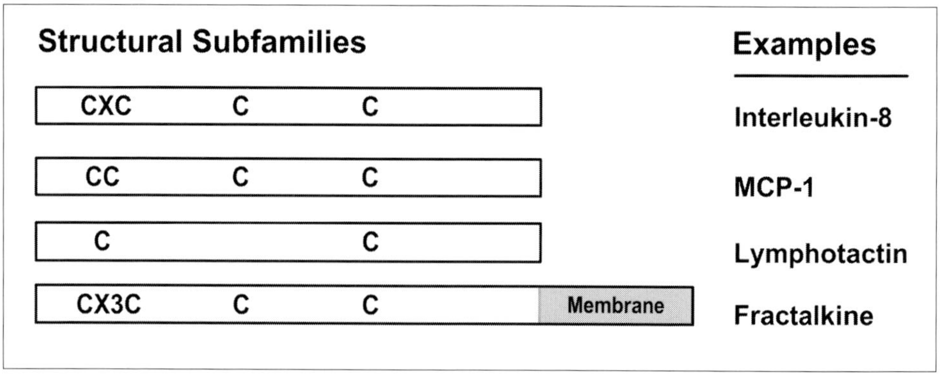

Figure 1
Chemokine subfamilies. The boxes represent the amino acid sequences, C indicates the position of cysteines that form intra-molecular disulphide bonds, and X stands for other amino acids. For each subfamily one representative example is named.

nists and even chemokine-binding proteins to gain control of leukocyte traffic. Viruses also learned to use chemokine receptors to infect cells [10].

The field moved in unexpected directions eventually showing that chemokines are involved in lymphocyte homing and in the house-keeping traffic that maintains the immune system effective. Roles for chemokines have also been suggested in haematopoiesis, morphogenesis, metastasis formation and angiogenesis. It has been shown that chemokine antagonists have anti-inflammatory and HIV-suppressing activity, and the development of low molecular weight antagonists has given rise to a major industrial effort toward therapy. The issue of targeting chemokines for therapeutic purposes is amply treated in Volume II of the present work.

Chemokine basics

Chemokines consist of approximately 70–130 amino acids including four conserved cysteines [11, 12]. As secretory proteins, they are synthesised with a leader sequence of 20–25 amino acids, which is cleaved off before release. Two main subfamilies, CXC and CC chemokines, are distinguished according to the position of the first two cysteines, which are separated by one amino acid (CXC) or adjacent (CC) (Fig. 1) [11, 12]. Two disulphide bonds, linking Cys1 to Cys3 and Cys2 to Cys4, confer to the chemokines their characteristic three-dimensional structure with a rigid core. The amino-terminal domain is short (3–10 amino acids) and structurally disordered, while the carboxyl-terminal helix consists of 20–60 amino acids. All

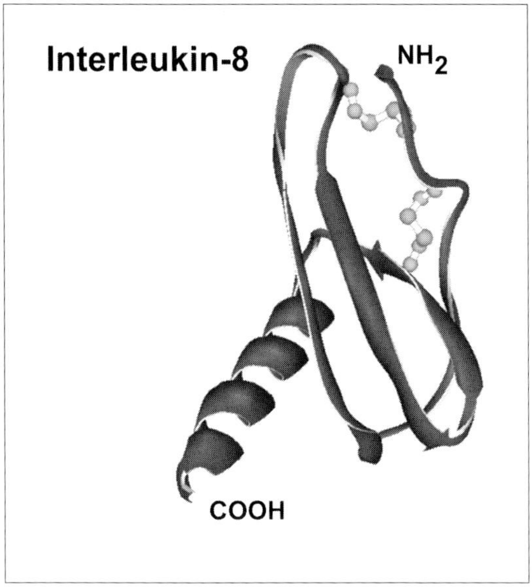

Figure 2
Three-dimensional structure of IL-8. In solution, all chemokines fold in this manner. The following, functionally relevant domains are visible: The receptor recognition (docking) region located within the exposed loop after the second cysteine, the receptor triggering region corresponding to the short amino-terminal sequence (NH_2), the prominent core consisting of three anti-parallel β-strands connected by loops, and a carboxyl-terminal α-helix (COOH). The characteristic disulphide bonds keep chemokines in their biologically active conformation.

chemokines are folded in this manner (Fig. 2) [13]. Few variants of the chemokine structure paradigm have been described. Lymphotactin has two, instead of four, conserved cysteines [14, 15], while fractalkine and CXCL16 are membrane-bound and have three and two amino acids, respectively, between the first two cysteines [14, 16–18]. The biological significance of these variants is largely unknown, but the adhesive properties of membrane-anchored chemokines may be relevant for leukocyte extravasation [19, 20].

Two chemokine nomenclature systems are used: the traditional abbreviations, such as IL-8 and MCP-1, which date back to the time of chemokine discovery, and a systematic nomenclature based on the structural motifs CXC, CC, XC, CX3C or CX2C, followed by 'L' (for ligand) and the number of the respective gene, e.g., CXCL8 for IL-8, CCL2 for MCP-1. The most common original names, together with the systematic designations, are presented in Table 1, and a complete listing with the most recent updates can be found at http://cytokine.medic.kumamoto-u.ac.jp. Chemokine receptors are designated according to the type of chemokine(s)

Table 1 - Human chemokines divided into CXC, CC, C, and CX3C subfamilies

Systematic [1]	Classical [2]	
CXC Chemokines		
CXCL1-3	GRO α, β, γ	Growth-related proteins α, β, γ
CXCL5	ENA-78	Epithelial cell-derived neutrophil-activating peptide 78
CXCL6	GCP-2	Granulocyte chemotactic protein 2
CXCL7	NAP-2	Neutrophil-activating peptide 2
CXCL8	IL-8	Interleukin 8
CXCL9	Mig	Monocyte/macrophage-activating, IFN-γ-inducible protein
CXCL10	IP10	IFN-γ-inducible 10 kDa protein
CXCL11	I-TAC	IFN-γ-inducible T cell alpha chemoattractant
CXCL12	SDF-1	Stromal cell-derived factor 1
CXCL13	BCA-1	B cell-attracting chemokine 1
CXCL14	BRAK	Breast and kidney-expressed chemokine
CC Chemokines		
CCL1	I-309	Intercrine-β glycoprotein 309
CCL2	MCP-1	Monocyte chemoattractant protein 1
CCL3,-4	MIP-1α, β	Macrophage inflammatory proteins 1α,1β
CCL5	RANTES	Regulated on activation, normal T cell expressed and secreted
CCL7	MCP-2	
CCL8	MCP-3	
CCL11	Eotaxin	Eosinophil chemoattractant protein

Systematic	Classical	
(CC chemokines continued)		
CCL13	MCP-4	
CCL14-16	HCC-1, 2, 4	Hemofiltrate CC chemokines 1, 2, 4
CCL17	TARC	Thymus and activation-regulated chemokine
CCL18	DC-CK1	Dendritic cell-derived CC chemokine 1
CCL19	ELC	EBI1(CCR7)-ligand chemokine
CCL20	LARC	Liver and activation-regulated chemokine
CCL21	SLC	Secondary lymphoid tissue chemokine
CCL22	MDC	Macrophage-derived chemokine
CCL23	MPIF-1	Myeloid progenitor inhibitory factor-1
CCL24	Eotaxin 2	
CCL25	TECK	Thymus-expressed chemokine
CCL26	Eotaxin 3	
CCL27	CTACK	Cutaneous T cell-attracting chemokine
CCL28	MEC	Mammary enriched chemokine
C Chemokines		
XCL1	Lymphotactin	(SCM-1α/single cysteine motif 1α)
XCL2	SCM-1β	Single cysteine motif 1β
CX3C Chemokine		
CX3CL1	fractalkine	

[1] *Systematic nomenclature is further defined at http://cytokine.medic.kumamoto-u.ac.jp.*
[2] *One representative out of several classical designations is listed for each chemokine.*

they bind (CXC, CC, XC, CX3C), followed by 'R' (for receptor) and a number reflecting the order of discovery.

Chemokines act via seven-trans-membrane domain receptors coupled to GTP-binding proteins. Most receptors recognise more than one chemokine and several chemokines bind to more than one receptor [21]. Structure–activity relationship studies have shown that CXC and CC chemokines have two sites of interaction with their receptors, one in the amino-terminal domain and the other within the exposed loop following the second cysteine. Both sites are kept in close proximity by the disulphide bonds. The loop region, which is conformationally rigid, appears to interact first and to function as a receptor-docking domain. This interaction restricts the mobility of the chemokine and presumably facilitates the binding of the amino-terminal domain that triggers a response (Fig. 3). All chemokines signal via receptors that are coupled to GTP-binding proteins of the Gi type and are sensitive to *B. pertussis* toxin. The signalling cascade induced by chemokines is typical for this class of seven-trans-membrane domain receptors [22].

Within the tissues, chemokines bind to glycosaminoglycans on the surface of cells and in the extracellular matrix by ionic interaction with basic residues in the core region and/or the carboxyl-terminal helix (Fig. 3) [23, 24]. Bound chemokines retain their full chemotactic activity and remain confined to the site where they are produced and released [25, 26]. This property explains the long-lasting, locally focused response to chemokines.

Receptor expression and chemokine driven leukocyte traffic regulation

In terms of function it is useful to differentiate between inflammatory and homeostatic chemokines. Inflammatory chemokines assure the recruitment of defence cells to sites of infection, tissue injury, inflammation and other disturbances of homeostasis. They are produced by a wide variety of tissue cells and by immigrating leukocytes at sites of pathological changes, act on receptors with broad selectivity, such as CXCR1, CXCR2, CXCR3, CCR1, CCR2, CCR3 and CCR5, and attract granulocytes, monocytes and lymphocytes. Homeostatic chemokines control the traffic of lymphocytes and their precursors during haematopoiesis in the bone marrow, the lymphoid and certain non-lymphoid tissues. They are expressed constitutively at homing sites within healthy tissues and act on receptors of high selectivity, which recognise a single, or at the most two, chemokines.

Initially chemokines were perceived as mediators of effector cell responses and the study of receptor expression was largely confined to phagocytes. Blood phagocytes express different sets of chemokine receptors. CXCR1 and CXCR2, the receptors or CXCL8/IL-8, are characteristic for neutrophils. Monocytes express CCR1, CCR2 and CCR5, eosinophils CCR1 and CCR3, while basophils express CCR1, CCR2 and CCR3. These patterns of receptors are characteristic for the different

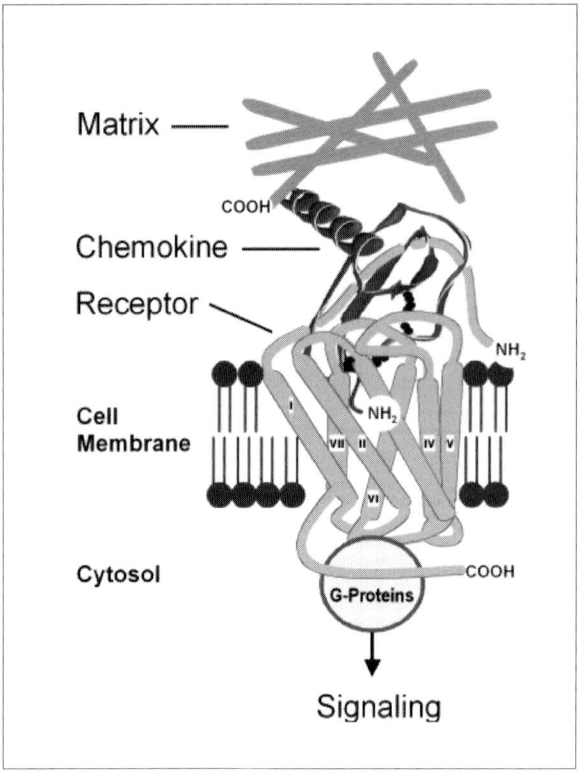

Figure 3
Interaction of chemokines with seven-trans-membrane domain receptors. The scheme shows the chemokine interacting with the receptor through its amino-terminal region and with extracellular glycosaminoglycans through heparin-binding regions, which are mostly localized in the carboxyl-terminal region (COOH). The chemokine-triggered receptor initiates the signaling cascade by activating a G-protein.

types of phagocytes and are sufficiently different to explain the selective recruitment of a single type of phagocyte, for instance, eosinophils in allergic inflammation or monocytes in chronic infectious lesions [27].

The results of studies on the responses of blood lymphocytes to chemokines were highly controversial until it was realised, that in these cells the expression of chemokine receptors changes considerably in dependence of differentiation and functional specialisation. It was first observed that culturing blood T cells in the presence of IL-2 progressively increases the expression of several receptors for inflammatory chemokines, such as CCR1, CCR2, CCR5 and CXCR3, and the chemotactic response to the respective ligands, e.g., CCL2/MCP-1, CCL3/MIP-1α,

CCL5/RANTES and CXCL10/IP10 [28]. The effect of IL-2 is reversible: Receptor numbers and responsiveness rapidly decline when the cytokine is withdrawn and are fully restored when it is supplied again. These observations indicated that chemokine receptor expression could be used to define different stages of T cell differentiation and the acquisition of particular functional properties.

Following up on these ideas, it was subsequently shown that Th1 and Th2 cells, as obtained by culturing in the presence of IL-2 and interferon-γ or IL-2 and IL-4, respectively, have different patterns of chemokine receptors: CCR5 and CXCR3 being characteristic for Th1 and CCR3 and CCR4 for Th2 cells [29, 30]. It was then shown that chemokine receptor detection by immunochemistry may be used for the identification of subtypes of T cells in tissues. Biopsies of rheumatoid synovium, which is rich in Th1 lymphocytes, stain strongly for CCR5, while a marked staining for CCR3 is detected at sites of allergic inflammation, where Th2 lymphocytes are recruited together with eosinophils [31].

CCR1, CCR2, CCR5 and CXCR3, the receptors that are up-regulated in T cells after treatment with IL-2, respond to inflammatory chemokines, which are induced at sites of infection and inflammation to recruit defence cells. When the T cells are stimulated with antibodies against CD3 and CD28, mimicking activation via the T cell receptor, they down-regulate the first set of receptors and up-regulate CCR7. A similar mechanism guides the traffic of dendritic cells. Inflammatory chemokines attract immature dendritic cells, expressing CCR1, CCR2 and CCR5, into inflamed tissues. The cells then mature, acquiring the capacity to capture and process antigens, and to present antigenic epitopes, and are thus ready to move on. CCR1, CCR2 and CCR5 are down-regulated and replaced by CCR7 and the mature dendritic cells migrate into the draining lymph nodes in response to CCL19/ELC and CCL21/SLC via CCR7 [32].

Effector and central memory T cells (TEM and TCM, respectively) can be distinguished according to their chemokine receptor outfit, which reflects their different role in a secondary immune response [33]. TEM cells have effector function. They produce IL-4 and interferon-γ, and may store perforin, and, owing to the absence of CCR7, can be recruited rapidly into inflamed tissues for immediate defence in response to inflammatory chemokines. By contrast, the CCR7-positive central memory T cells (TCM) have no immediate effector function. They represent a clonally expanded memory cell pool, are attracted to lymph nodes after a secondary antigen challenge, and can stimulate dendritic cells to produce IL-12, provide help to antigen-specific B cells, and generate a new wave of effector T cells [33].

Control of lymphocyte traffic in disease-unrelated processes

Homeostatic chemokines control the relocation and recirculation of lymphocytes in the context of maturation, differentiation and activation, and ensure their correct positioning within discrete areas of primary and secondary lymphoid organs [34, 35].

The recognition that chemokines direct the homeostatic traffic of lymphocytes goes back to the work by Lipp and colleagues [36] who found that the deletion of the gene of the putative chemokine receptor BLR1 (which was renamed CXCR5 after identification of its ligand chemokine, CXCL13/BCA-1 [37, 38]) impaired the formation of Peyer's patches and inguinal lymph modes because of the inability of CXCR5-deleted B cells to home into follicular areas. Subsequent work elucidated the role of another receptor for homeostatic chemokines, CCR7, which binds CCL19/ELC and CCL21/SLC [39]. Follicle formation in lymphoid tissues depends on immigration and settling of B and T cells. Both types of lymphocytes bear CCR7, they are recruited in response to CCL21/SLC expressed in high-endothelial venules and migrate to the parafollicular area in response to CCL19/ELC and CCL21/SLC. The B cells, which also bear CXCR5, are attracted into the follicles, where CXCL13/BCA-1 is expressed.

It was subsequently found that T cells acquire CXCR5 on activation, in particular on contact with antigen-presenting dendritic cells. Such cells can thus enter the follicles in response to CXCL13/BCA-1 and fulfil a helper function to B cells by enhancing antibody production. Some re-enter circulation as a small pool of memory cells [40, 41]. CXCR5-bearing T cells represent a novel type of effectors. They differ from Th1 and Th2 cells as they markedly enhance antibody production when co-cultured with B cells and do not express cytokines that are characteristic of Th1 or Th2 cells [42]. Owing to their follicular homing properties and function, these cells are called follicular B helper T cells (TFH). The possible involvement of TFH cells in immune pathology, including autoimmune diseases with B cell involvement is presently under study.

Peripheral immune surveillance T cells

The skin, the gut and the lung are the main sites of pathogen entry into the body owing to their huge contact area to the outside. Immune defence in these tissues is assured by dedicated lymphoid structures (like the mucosa-associated lymphoid tissue of the lung and the gastrointestinal tract) and by a large population of resident T cells, which are distributed throughout the tissue. The mechanism of the tissue-specific entry of immune surveillance T cells is studied by searching for chemokines that are constitutively expressed by the endothelia of blood micro-vessels, the main site of leukocyte extravasation, and by determining the pattern of chemokine receptor expression of the resident T cells. In the skin, most T cells cluster around post-capillary venules of the superficial dermal plexus. *In situ* studies have shown that these cells express CCR8, and that CCL1/I-309, its only ligand, is produced constitutively in blood micro-vessels (as well as in Langerhans cells and melanocytes of healthy epidermis) but not in keratinocytes or fibroblasts [43]. No other chemokine and receptor combination appears to satisfy the requirements for constitutive expression, local distribution and selectivity. It is thus assumed that the homeostatic traffic of skin-homing T cells is based, at least in part, on the recruitment of circulating CCR8 expressing T cells in response to cutaneous CCL1/I-309 [44]. One would expect that similar mechanisms regulate the selective homing of T cells into the gut and the lung. It has been shown that effector T cells home into the small intestine in response to CCL25/TECK acting via its receptor, CCR9 [45, 46], but the role of CCR9 and its ligand chemokine in the homeostatic traffic of gut-selective T cells is still a matter of debate. The studies of the skin indicate that peripheral immune surveillance T cells (TPS), in contrast to TCM and TEM cells, fulfil a "first line of defence" function, like other sentinel cells, and it is thus reasonable to assume that TPS cells are present in other frontier tissues [44].

Volume I focuses on the functions of chemokines in immunobiology, as the title indicates, with particular attention to the control of T cell traffic in inflammation and homeostasis. In view of major recent progress, the properties of newly-defined T cell subsets with *bona fide* effector and/or memory functions, namely TCM, TEM and TPS cells will be discussed in relation to Th1 and Th2 cells. A special chapter is dedicated to NK cells and γδ T cells, which share certain features with effector T cells. Adaptive immunity, including immune homeostasis and antimicrobial defence, fully depends on antigen-presentation and co-stimulation by dendritic cells and, therefore, an update on the control of dendritic cell traffic by chemokines is presented. Chemokine-induced cellular responses are mediated by selective receptors. The complex molecular networks involving soluble and membrane-bound mediators that are activated on chemokine receptor triggering are considered in a separate chapter. Since considerable progress has been made recently in the study of the homeostatic functions of chemokines, the local, constitutive production of chemokines in the tissues, in particular by the endothelial cells of micro-vessels, and its role in leukocyte transendothelial migration has been given special consideration. A chapter considers the modification of chemokines and chemokine activities by proteases, as well as the phenomenon of inhibition or potentiation of chemokine-induced responses by other chemokines or chemokine derivatives. These interactions will eventually deepen our understanding of leukocyte recruiting in inflammation, when several chemokines are produced concomitantly. The last part of the volume is dedicated to chemokine-mediated responses that involve tissue cells and microbes. New insides are presented on the cross-talk between G-protein-coupled receptors on neurons and leukocytes, the influence of virus-encoded chemokines on the immune system of the host, the function of chemokine receptors in tissue cells, and the involvement of chemokines and related peptides in antimicrobial defence. The state-of-the-art view on chemokine immunobiology should provide the context for discussing pathology and therapy-related aspects of chemokine research, which are the main focus of Volume II.

References

1 Walz A, Peveri P, Aschauer H, Baggiolini M (1987) Purification and amino acid sequencing of NAF, a novel neutrophil-activating factor produced by monocytes. *Biochem Biophys Res Commun* 149: 755–761

2 Yoshimura T, Matsushima K, Tanaka S, Robinson EA, Appella E, Oppenheim JJ, Leonard EJ (1987) Purification of a human monocyte-derived neutrophil chemotactic factor that has peptide sequence similarity to other host defense cytokines. *Proc Natl Acad Sci USA* 84: 9233–9237

3 Schröder J-M, Mrowietz U, Morita E, Christophers E (1987) Purification and partial biochemical characterization of a human monocyte-derived, neutrophil-activating peptide that lacks interleukin 1 activity. *J Immunol* 139: 3474–3483

4 Van Damme J, Van Beeumen J, Opdenakker G, Billiau A (1988) A novel, NH2-terminal sequence-characterized human monokine possessing neutrophil chemotactic, skin-reactive, and granulocytosis-promoting activity. *J Exp Med* 167: 1364–1376

5 Thelen M, Peveri P, Kernen P, von Tscharner V, Walz A, Baggiolini M (1988) Mechanism of neutrophil activation by NAF, a novel monocyte-derived peptide agonist. *FASEB J* 2: 2702–2706

6 Baggiolini M, Walz A, Kunkel SL (1989) Neutrophil-activating peptide-1/interleukin 8, a novel cytokine that activates neutrophils. *J Clin Invest* 84: 1045–1049

7 Lindley I, Aschauer H, Seifert JM, Lam C, Brunowsky W, Kownatzki E, Thelen M, Peveri P, Dewald B, von Tscharner V et al (1988) Synthesis and expression in *Escherichia coli* of the gene encoding monocyte-derived neutrophil-activating factor: Biological equivalence between natural and recombinant neutrophil-activating factor. *Proc Natl Acad Sci USA* 85: 9199–9203

8 Clark-Lewis I, Moser B, Walz A, Baggiolini M, Scott GJ, Aebersold R (1991) Chemical synthesis, purification, and characterization of two inflammatory proteins, neutrophil activating peptide 1 (interleukin-8) and neutrophil activating peptide 2. *Biochemistry* 30: 3128–3135

9 Baggiolini M (2001) Chemokines in pathology and medicine. J Intern Med 250: 91–104

10 Murphy PM (2001) Viral exploitation and subversion of the immune system through chemokine mimicry. *Nat Immunol* 2: 116–122

11 Baggiolini M, Dewald B, Moser B (1994) Interleukin-8 and related chemotactic cytokines – CXC and CC chemokines. *Adv Immunol* 55: 97–179

12 Baggiolini M, Dewald B, Moser B (1997) Human chemokines: An update. *Annu Rev Immunol* 15: 675–705

13 Clark-Lewis I, Kim K-S, Rajarathnam K, Gong J-H, Dewald B, Moser B, Baggiolini M, Sykes BD (1995) Structure-activity relationships of chemokines. *J Leukocyte Biol* 57: 703–711

14 Kelner GS, Kennedy J, Bacon KB, Kleyensteuber S, Largaespada DA, Jenkins NA,

Copeland NG, Bazan JF, Moore KW, Schall TJ et al (1994) Lymphotactin: A cytokine that represents a new class of chemokine. *Science* 266: 1395–1399

15 Kennedy J, Kelner GS, Kleyensteuber S, Schall TJ, Weiss MC, Yssel H, Schneider PV, Cocks BG, Bacon KB, Zlotnik A (1995) Molecular cloning and functional characterization of human lymphotactin. *J Immunol* 155: 203–209

16 Bazan JF, Bacon KB, Hardiman G, Wang W, Soo K, Rossi D, Greaves DR, Zlotnik A, Schall TJ (1997) A new class of membrane-bound chemokine with a CX3C motif. *Nature* 385: 640–644

17 Pan Y, Lloyd C, Zhou H, Dolich S, Deeds J, Gonzalo JA, Vath J, Gosselin M, Ma JY, Dussault B et al (1997) Neurotactin, a membrane-anchored chemokine upregulated in brain inflammation. *Nature* 387: 611–617

18 Matloubian M, David A, Engel S, Ryan JE, Cyster JG (2000) A transmembrane CXC chemokine is a ligand for HIV-coreceptor Bonzo. *Nat Immunol* 1: 298–304

19 Imai T, Hieshima K, Haskell C, Baba M, Nagira M, Nishimura M, Kakizaki M, Takagi S, Nomiyama H, Schall TJ et al (1997) Identification and molecular characterization of fractalkine receptor CX3CR1, which mediates both leukocyte migration and adhesion. *Cell* 91: 521–530

20 Fong AM, Robinson LA, Steeber DA, Tedder TF, Yoshie O, Imai T, Patel DD (1998) Fractalkine and CX3CR1 mediate a novel mechanism of leukocyte capture, firm adhesion, and activation under physiologic flow. *J Exp Med* 188: 1413–1419

21 Murphy PM, Baggiolini M, Charo IF, Hebert CA, Horuk R, Matsushima K, Miller LH, Oppenheim JJ, Power CA (2000) International union of pharmacology. XXII. Nomenclature for chemokine receptors. *Pharmacol Rev* 52: 145–176

22 Thelen M (2001) Dancing to the tune of chemokines. *Nat Immunol* 2: 129–134

23 Chakravarty L, Rogers L, Quach T, Breckenridge S, Kolattukudy PE (1998) Lysine 58 and histidine 66 at the C-terminal α-helix of monocyte chemoattractant protein-1 are essential for glycosaminoglycan binding. *J Biol Chem* 273: 29641–29647

24 Amara A, Lorthioir O, Valenzuela A, Magerus A, Thelen M, Montes M, Virelizier JL, Delepierre M, Baleux F, Lortat-Jacob H et al (1999) Stromal cell-derived factor-1a associates with heparan sulfates through the first b-strand of the chemokine. *J Biol Chem* 274: 23916–23925

25 Webb LMC, Ehrengruber MU, Clark-Lewis I, Baggiolini M, Rot A (1993) Binding to heparan sulfate or heparin enhances neutrophil responses to interleukin 8. *Proc Natl Acad Sci USA* 90: 7158–7162

26 Middleton J, Neil S, Wintle J, Clark-Lewis I, Moore H, Lam C, Auer M, Hub E, Rot A (1997) Transcytosis and surface presentation of IL-8 ky venular endothelial cells. *Cell* 91: 385–395

27 Baggiolini M, Dahinden CA (1994) CC chemokines in allergic inflammation. *Immunol Today* 15: 127–133

28 Loetscher P, Seitz M, Baggiolini M, Moser B (1996) Interleukin-2 regulates CC chemokine receptor expression and chemotactic responsiveness in T lymphocytes. *J Exp Med* 184: 569–577

29 Bonecchi R, Bianchi G, Bordignon PP, D'Ambrosio D, Lang R, Borsatti A, Sozzani S, Allavena P, Gray PA, Mantovani A et al (1998) Differential expression of chemokine receptors and chemotactic responsiveness of type 1 T helper cells (Th1s) and Th2s. *J Exp Med* 187: 129–134

30 Sallusto F, Lenig D, Mackay CR, Lanzavecchia A (1998) Flexible programs of chemokine receptor expression on human polarized T helper 1 and 2 lymphocytes. *J Exp Med* 187: 875–883

31 Loetscher P, Moser B, Baggiolini M (2000) Chemokines and their receptors in lymphocyte traffic and HIV infection. *Adv Immunol* 74: 127–180

32 Sallusto F, Mackay CR, Lanzavecchia A (2000) The role of chemokine receptors in primary, effector, and memory immune responses. *Annu Rev Immunol* 18: 593–620

33 Sallusto F, Geginat J, Lanzavecchia A (2004) Central memory and effector memory T cell subsets: function, generation, and maintenance. *Annu Rev Immunol* 22: 745–763

34 Moser B, Loetscher P (2001) Lymphocyte traffic control by chemokines. *Nat Immunol* 2: 123–128

35 Moser B, Wolf M, Walz A, Loetscher P (2004) Chemokines: multiple levels of leukocyte migration control. *Trends Immunol* 25: 75–84

36 Förster R, Mattis AE, Kremmer E, Wolf E, Brem G, Lipp M (1996) A putative chemokine receptor, BLR1, directs B cell migration to defined lymphoid organs and specific anatomic compartments of the spleen. *Cell* 87: 1037–1047

37 Legler DF, Loetscher M, Roos RS, Clark-Lewis I, Baggiolini M, Moser B (1998) B cell-attracting chemokine 1, a human CXC chemokine expressed in lymphoid tissues, selectively attracts B lymphocytes via BLR1/CXCR5. *J Exp Med* 187: 655–660

38 Gunn MD, Ngo VN, Ansel KM, Ekland EH, Cyster JG, Williams LT (1998) A B-cell-homing chemokine made in lymphoid follicles activates Burkitt's lymphoma receptor-1. *Nature* 391: 799–803

39 Förster R, Schubel A, Breitfeld D, Kremmer E, Renner-Müller I, Wolf E, Lipp M (1999) CCR7 coordinates the primary immune response by establishing functional microenvironments in secondary lymphoid organs. *Cell* 99: 23–33

40 Schaerli P, Willimann K, Lang AB, Lipp M, Loetscher P, Moser B (2000) CXC chemokine receptor 5 expression defines follicular homing T cells with B cell helper function. *J Exp Med* 192: 1553–1562

41 Breitfeld D, Ohl L, Kremmer E, Ellwart J, Sallusto F, Lipp M, Forster R (2000) Follicular B helper T cells express CXC chemokine receptor 5, localize to B cell follicles, and support immunoglobulin production. *J Exp Med* 192: 1545–1552

42 Moser B, Schaerli P, Loetscher P (2002) CXCR5(+) T cells: follicular homing takes center stage in T-helper-cell responses. *Trends Immunol* 23: 250–254

43 Schaerli P, Ebert L, Willimann K, Blaser A, Roos RS, Loetscher P, Moser B (2004) A skin-selective homing mechanism for human immune surveillance T cells. *J Exp Med* 199: 1265–1275

44 Ebert LM, Schaerli P, Moser B (2005) Chemokine-mediated control of T cell traffic in lymphoid and peripheral tissues. *Mol Immunol* 42: 799–809

45 Svensson M, Marsal J, Ericsson A, Carramolino L, Broden T, Marquez G, Agace WW (2002) CCL25 mediates the localization of recently activated CD8alphabeta(+) lymphocytes to the small-intestinal mucosa. *J Clin Invest* 110: 1113–1121
46 Johansson-Lindbom B, Svensson M, Wurbel MA, Malissen B, Marquez G, Agace W (2003) Selective generation of gut tropic T cells in gut-associated lymphoid tissue (GALT): requirement for GALT dendritic cells and adjuvant. *J Exp Med* 198: 963–969

Cellular targets in innate and adaptive immunity

Traffic of T lymphocytes

Charles R. Mackay[1] *and Bernhard Moser*[2]

[1]Garvan Institute of Medical Research, Darlinghurst NSW 2010, Australia; [2]Institute of Cell Biology, University of Bern, Baltzerstrasse 4, 3012 Bern, Switzerland

Introduction

A large number of chemokines are involved in the control of T cell migration, which may reflect the multitude of distinct T cell subsets participating in immune processes at various locations throughout the body. In our discussion it may be helpful to divide the chemokines into two functional subfamilies, termed homeostatic and inflammatory chemokines [1, 2]. Homeostatic chemokines navigate leukocytes during haematopoiesis in the bone marrow and thymus, during initiation of adaptive immune responses in the spleen and lymph nodes (LNs), and during immune surveillance of healthy peripheral tissues. Inflammatory chemokines, by contrast, control the recruitment of effector leukocytes in infection, inflammation, tissue injury and tumours. This classification is not strict since "dual-function" chemokines may also exist [1].

Chemokines present on vascular endothelia control leukocyte extravasation, as discussed in detail in Chapter 6, Vol. I; whereas chemokines produced by tissue cells control the homing of responding leukocytes to distinct tissue locations. We wish to emphasise that the migration properties and function represent two sides of the same coin. Therefore, detailed examination of the type and regulation of chemokine receptors present on a particular subset of T cells provides invaluable information about their physiological role. Table 1 represents a list in progress of T cell subsets defined by their migratory potential. This view extends the classical approach in immunological research dealing with "endpoint" analyses, i.e., analyses of *in vitro* cultured T cells or of T cells recovered from laboratory animals after *in vivo* manipulations. The following discussion summarises our current knowledge about chemokines involved in traffic control related to the initiation of αβ T cell responses and effector/memory functions. Those chemokines acting on γδ T cells and T cell precursors are reviewed in great detail in the chapter by Jin and Morita.

Chemokine Biology – Basic Research and Clinical Application, Volume I
edited by Bernhard Moser, Gordon L. Letts and Kuldeep Neote

Table 1 - T cell subsets defined by migration properties

T cell subset	ChemRs[a]	Residence[b]	Phenotype[c]
Naïve T	CCR7 (CXCR4)	Blood LNs, PPs, Spleen	CD45RA+ (CD45RO−), non-differentiated, resting
T_{FH} (follicular B helper)	CXCR5 (CXCR4, CCR7)	LNs, PPs, Spleen (Blood)	CD45RO+ CD4+, non-differentiated (but ICOS+ and IL-10 secretion), activated, transient
Effector T	Inflammatory ChemRs	Inflammation[d]	CD45RO+, differentiated, (cytokine secretion, target cell lysis), activated, short-lived
T_{CM} (central memory)	CCR7 (CXCR4)	Blood LNs, PPs, Spleen	CD45RO+, non/partial-differentiated, resting, long-lived
T_{EM} (effector memory)	Inflammatory ChemRs	Blood Inflammation	CD45RO+, differentiated, (cytokine secretion, target cell lysis), resting, long-lived
T_{PS} (peripheral immune surveillance)	Homeostatic ChemRs	Healthy peripheral tissues	CD45RO+ (partial CD45RA+), differentiated, partial-activated, long-lived

[a]ChemRs, chemokine receptors; chemokine receptors in brackets are of secondary importance.
[b]Residence refers to the primary location within the body of the respective T cell subset.
[c]Phenotype refers to memory status, longevity, and cellular responses defined by cell surface markers and TCR-triggered effector functions.
[d]Inflammation stands for all sites where inflammatory chemokines are being produced, including acute and chronic infections, autoimmune diseases and tumours.

Initiation of adaptive immune responses

As part of their normal route of recirculation, naïve T cells regularly leave the blood and enter LNs by passing through high endothelial venules (HEVs) [3, 4] (Fig. 1). Passage of T cells through the HEV barrier underlies the same paradigm that applies to any other blood endothelia: leukocyte rolling, chemokine-mediated activation and subsequent firm adhesion, followed by leukocyte transendothelial migration [4, 5]. Chemokines play a decisive role in controlling the type of cells allowed to enter this site. Here, the two homeostatic chemokines, CCL19 and CCL21, have been shown to be essential in the transmigration of HEVs. The shared receptor for these two chemokines, CCR7, is uniformly expressed by all naïve T cells, as well as a sub-

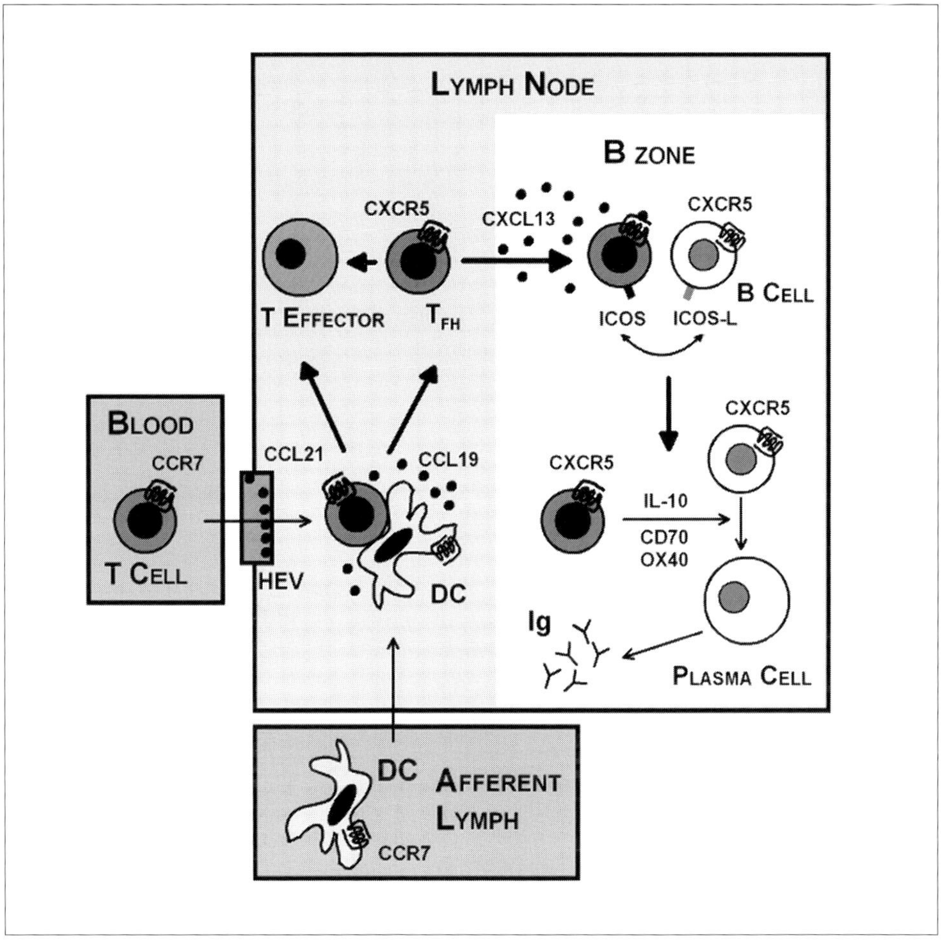

Figure 1

Chemokines in the control of primary T cell responses

The expression of CCR7 by naïve T cells and T_{CM} allows entry into the LNs via HEVs, and subsequent co-localisation with CCR7-expressing DCs in the T zone. T cell priming is mediated by antigen-loaded DCs and results in the generation of effector T cells and CXCR5-expressing T_{FH} cells. Expression of CXCR5 by T_{FH} cells makes them responsive to the chemokine CXCL13 produced by cells within B cell follicles, resulting in re-localisation of T_{FH} cells to the B cell compartment. Subsequent interactions between B cells and T_{FH} cells, possibly involving the interaction of ICOS on T_{FH} cells with the ICOS-ligand on B cells, induce T cell differentiation (including increased expression of CD70 and OX40 and enhanced IL-10 secretion). In return, TFH cells provide help to B cells for plasma cell differentiation and antibody production. T Effector, effector T cell; T_{FH}, follicular B helper T cell; HEV, high endothelial venule; DC, dendritic cell.

set of resting memory T cells known as central memory T (T_{CM}) cells (Tab. 1) [6]. Both chemokines are displayed within the lumen of HEVs; CCL21 is constitutively expressed by HEVs [7, 8], while CCL19 is produced by other cells within the LN but becomes displayed on the HEV lumen following transcytosis across the endothelial barrier [9]. The importance of CCR7 and its ligands in T cell entrance to LNs has been clearly demonstrated in studies of mice genetically deficient in CCR7 or mice harbouring a spontaneous mutation (*plt*) that results in defective production of CCL19 and CCL21 [10]. These mutant mouse strains show greatly reduced numbers of naïve T cells in LNs, which is due to their inability to firmly adhere to, and transmigrate across, HEVs. CCL19 also orchestrates the co-localisation of freshly recruited T cells with dendritic cells (DCs) in the T zone, and this process is of fundamental importance for antigen-presentation to T cells and induction of primary immune responses (Fig. 1) [11, 12].

Circulating naïve T cells express, in addition to CCR7, only few other chemokine receptors, which explains their broad exclusion from healthy peripheral tissues and acute inflammatory diseases [1]. All naïve T cells uniformly express CXCR4 whereas only minor fractions are positive for CCR8 and CCR9. Recent findings support a role for CCR8 and its single ligand CCL1 in the control of peripheral immune surveillance T (T_{PS}) cells within normal human skin (Tab. 1) [13]. These cutaneous CCR8$^+$ T_{PS} cells are antigen-experienced, partially activated Th1/Tc1 cells that may contribute to the local inflammatory cascade at the site of pathogen entry (see below). A second memory T cell subset includes the very few CCR8$^+$ T cells present in peripheral blood, which may be related to regulatory T cells generated during thymocyte development [13, 14]. Also, thymocytes frequently express this chemokine receptor. As for CCR8, a role for CCR9 in the entry of T cells into LNs has not been demonstrated. The single ligand CCL25 for CCR9 is selectively expressed in small intestine, raising the possibility that this chemokine is involved in the traffic of naïve CCR9$^+$ T cells to small intestinal lymphoid structures [15]. However, it is also possible that expression of CCR9 by naïve T cells is simply a remnant of their development in the thymus, as CCR9 is broadly expressed by thymocytes and the CCR9$^+$ subset of naïve T cells declines with age or surgical thymectomy [15, 16]. By contrast, CXCR4 does appear to contribute to T cell entry into LNs. Although greatly reduced in number, some T cells still enter the LNs in CCL19/CCL21-deficient (*plt*) mice, and this residual migration is completely ablated when CXCR4-deficient T cells were adoptively transferred [17]. Furthermore, CXCL12 (the ligand for CXCR4) has been shown to promote transendothelial migration of T cells across the HEVs [18]. Hence, CCR7 and CXCR4 may co-operate in the task of naïve T cell recruitment into LNs. Collectively, naïve T cells are not only "naïve" in terms of antigen experience but also in terms of migration behaviour, which controls their continuous recirculation between blood and secondary lymphoid tissues. This is in clear contrast to effector/memory T cells characterised by a multiple receptors for inflammatory chemokines, by single or repeated exposure to antigen and by TCR-triggered effector functions (see below).

Effector T cell generation is a highly sophisticated process that depends on multiple and partially overlapping steps, including T cell priming, proliferation and development of effector functions [19]. Of particular importance to the present discussion, priming of $CD4^+$ T cells results in the novel expression of CXCR5 [20, 21], a chemokine receptor otherwise broadly expressed on resting B cells [22, 23] (Fig. 1). CXCL13, the ligand for CXCR5, is markedly produced within the B cell-rich follicular compartment of secondary lymphoid tissues, but is absent from the adjacent T zone [23, 24]. LN and Peyer's patches (PP) neogenesis largely depend on this chemokine system [25], and the architecture of the follicular compartments within spleen and LNs are greatly disturbed in CXCR5-deficient mice [26], supporting the notion that CXCL13 and its receptor are essential contributors to follicular activities.

In clear contrast to the T zone chemokines CCL19 and CL21, the single CXCR5 ligand CXCL13 is selectively produced within the B cell compartment, suggesting that the acquisition of CXCR5 by recently primed $CD4^+$ T cells would drive their relocation to the B cell follicles (Fig. 1). Indeed, several studies have documented a temporary relocation of T cells to the outer edge of the follicles in response to immunisation [27, 28]. In mice, follicular migration of primed T cells occurs rapidly after immune response initiation, well before the generation of effector T cells [21, 28], and this observation fully agrees with the kinetics of CXCR5 expression on human T cells. Peak levels of CXCR5 are acquired within the first 2–3 days of *in vitro* stimulation of naïve human T cells, well before induction of T cell polarisation, as assessed by the absence of effector functions (target cell lysis, cytokine production) [20, 24, 29, 30]. CXCR5 is rapidly lost, however, during *in vitro* T cell proliferation. Also, it is not possible to generate T cell lines stably expressing CXCR5, suggesting that maintenance of this chemokine receptor relies on a particular microenvironment (see below).

$CXCR5^+$ T cells can provide potent help for antibody production during co-culture with B cells, and this characteristic together with the follicular homing behaviour prompted their designation as follicular B helper T (T_{FH}) cells [24, 29]. Of interest, the majority of CD4+ T cells appear to rapidly express CXCR5 upon stimulation, whereas CXCR5 is very infrequent on $CD8^+$ T cells [24], supporting the notion that T_{FH} cells contribute to B cell responses (Fig. 1). The mechanism by which T_{FH} cells provide help to B cells is a subject of current investigations. Except for IL-2, T_{FH} cells from tonsils are poor cytokine producers [24, 29], suggesting that newly generated T_{FH} cells require further differentiation in order to become effective helpers for plasma cell differentiation and antibody production.

The ability to provide effective B cell help is one important aspect of T_{FH} cell differentiation that may be controlled by B cells (Fig. 1). This concept is consistent with *in vivo* studies suggesting that B cells 'solicit their own help' from the T cell compartment. Moreover, recent studies in our laboratory directly demonstrate that B cells can indeed influence the phenotype in T_{FH} cells during co-culture [31]. Of interest, tonsillar as well as *in vitro* generated T_{FH} cells strongly express ICOS, a recently identified co-stimulatory molecule with critical functions in T helper and B cell responses [31, 32]. Newly generated T_{FH} cells express a phenotype consistent with induction of B cell proliferation. However, during co-culture with B cells these cells assume a B helper phenotype characterised by loss of CD154, induction of CD70 and an increase in IL-10 production. Also, B cells help to preserve a LN migration phenotype in proliferating T_{FH} cells, thus, directly preventing their premature exit out of LNs. It will be interesting to see if follicular T_{FH} cells shuttle back and forth between follicular compartment and T zone and if this steady relocation contributes to T helper cell differentiation.

Effector *versus* memory T cell traffic

Immunological memory resides within the subset of previously activated T cells. These T cells for the most part express the exon A-deficient (RO) isoform of CD45, as well as various other markers. One important distinction between naïve T cells and the various subsets of previously activated T cells is the expression of homing molecules such as selectins, integrins and chemokine receptors.

Th1 and Th2 T cells

Naïve T cells differentiate to effector cells in lymphoid organs, such as spleen, LNs and Peyer's patches (PPs). However, the principal sites where T helper (Th) cells and cytotoxic T cells exert their function are peripheral tissues, where pathogens are frequently encountered. Thus, effector cells up-regulate receptors for inflammation-induced endothelial adhesion molecules and inflammatory chemokines [33, 34]. Different pathogens require different effector responses, produced upon antigen-recognition by distinct T cell subsets. For instance, the T helper subsets Th1 and Th2 cells secrete non overlapping sets of cytokines (INF γ *versus* IL 4, IL 5 and IL 13), neutralise distinct types of pathogens (intracellular *versus* extracellular), express characteristic chemoattractant receptors and obey different traffic signals [35, 36]. Distinctive chemokine receptors on Th1 cells include CCR5 and CXCR3 [37, 38], which bind inflammatory chemokines. In rheumatoid arthritis (RA) and multiple sclerosis (MS), both often thought of as Th1-related, virtually all infiltrating T cells express CCR5 and CXCR3 [39, 40]. People with a homozygous mutation that disrupts the CCR5 gene may also be less susceptible to some inflammatory disorders, including RA [41, 42]. Adhesion molecules also play a role; Th1 cells express abundant selectin ligands. P- and E-selectin, which are up-regulated on inflamed endothelium, and their ligand, P-selectin glycoprotein ligand 1 (PSGL-1), are critical for Th1 cell migration to inflamed skin [43, 44] and peritoneum [45]. Expression of fucosyltransferase-VII is necessary for cells to synthesise selectin ligands [46]. This enzyme is induced by IL-12, which drives Th1 differentiation, whereas T cell exposure to the Th2 cytokine IL-4 down-modulates selectin ligand expression [47, 48].

Th2 cells also express distinctive chemoattractant receptors, including CRTh2 and CCR3 [49–51]. Eotaxin, a ligand of CCR3, has been implicated in eosinophil recruitment into hyper-reactive airways and is prominent in mucosal tissues undergoing allergic and anti-parasitic responses [52]. Eotaxin production is stimulated by Th2 cytokines, such as IL-4 or IL-13, and is absent from Th1-mediated lesions [53]. CCR3 is also expressed on basophils and mast cells, which presumably allows these allergy-related leukocytes to co-localise with Th2 cells and support local allergic inflammation. Other chemoattractant receptors that were originally identified as Th2-associated included CCR4, CCR8 and CXCR4; however, some of these associations are not holding up, or do not appear to be relevant *in vivo*. For instance, CCR8-deficient mice were originally shown to have defective Th2-type responses [54], but subsequent reports have failed to support these findings [55, 56]. The true physiological function of CCR8 is more likely related to skin-homing [13], since the majority of human T cells in healthy (non-inflamed) skin express CCR8, and interestingly these T cells display a Th1 rather than a Th2 cytokine profile [13]. CCL1, the only ligand of human CCR8, is constitutively expressed in skin, notably in dermal microvessels and epidermal antigen presenting cells (APCs) [13]; hence, this chemokine system may function in homeostatic T cell traffic through normal skin. Similarly, CCL17, a ligand of CCR4, is expressed in non-inflamed dermal microvessels, and may also direct homeostatic T cell traffic through skin [57]. CCR4 has been identified as a skin-homing receptor for memory T cells [57a], and Th2 memory cells derived from skin lesions of atopic dermatitis patients selectively migrated to human skin grafts transplanted onto severe combined immune deficiency (SCID) mice in response to ligands for CCR4 but not to ligands for CCR3, CCR8 or CXCR3 [58]. Future studies will tell if these chemokine systems fulfil a major role in maintaining local memory T cell traffic under homeostatic (non-inflamed) conditions or whether they recruit effector/memory T cells as a consequence of local inflammation.

T_{CM} cells *versus* T_{EM} cells

The most obvious and abundant cell surface marker for circulating memory T cells is CD45RO, which is rapidly induced upon T cell receptor triggering in naïve $CD45RA^+$ (but $CD45RO^-$) T cells and which is maintained throughout the lifespan of antigen-experienced T cells. Consequently, "memory" T cells are highly heterogeneous in their expression of homing related molecules, such as adhesion molecules and receptors for inflammatory or homeostatic chemokines [1, 2]. In a highly cited study, Sallusto and Lanzavecchia [6] identified two major subsets of memory cells in human peripheral blood based on the expression of CCR7. This chemokine receptor divides memory T cells into CCR7$^+$ central memory (T_{CM}) cells and CCR7– effector memory (T_{EM}) cells. Most blood T_{CM} cells also express L-selectin, which, together with CCR7, defines a LN-homing phenotype. Conversely, T_{EM} cells express homing receptors for peripheral tissues and display characteristic features of effector T cells upon TCR activation. T_{CM} cells do not exert immediate effector function when stimulated with antigen, i.e., are not thought to become engaged in antimicrobial responses within infected tissues, but instead may participate in recall (or memory) responses that are initiated in secondary lymphoid tissues. It is likely that immunological memory is contained in both of these subsets; however the relative importance of each subset is not yet known. Consequently, the prevailing view is that T_{CM} cells and T_{EM} cells differ in the location where recall antigens are encountered, which in broad terms include spleen, LNs and PPs for T_{CM} cells and peripheral organs, in particular the skin and mucosal tissues of the airways and digestive tract, for T_{EM} cells. We have performed Affymetrix Genechip analyses on T_{CM} cells and T_{EM} cells isolated from human peripheral blood, and found surprisingly few genes that were differentially expressed between the two memory subsets, other than CCR7 and L-selectin, the markers used to sort these subsets (Chtanova and Mackay, unpublished).

Regulatory T cells

Regulatory T (T_{reg}) cells are now widely accepted as an effector T cell type that serves to subdue immune responses, thereby providing a level of tolerance in peripheral immunity. T_{reg} cells express CD4, CD25 and Foxp3, and through expression of factors such as IL-10 and/or TGF-β exert a negative influence on T cell activation [59]. The actual sites where Treg cells fulfil their inhibitory function, and the diversity in T_{reg} cells with regard to site of generation and traffic pattern, are interesting topics of current investigations. For instance, a pulmonary T_{reg} subset has been described that may function specifically in the mucosal tissue of the airways [60]. Moreover, heterogeneous expression of various homing molecules on T_{reg} cells is consistent with their ability to suppress immune responses at various locations. In mice, L-selectin and CCR7 are expressed on a subset of LN-homing T_{reg} cells that inhibit diabetes induced by islet-infiltrating T cells [60a]. Migration to secondary lymphoid organs may be necessary for the antigen-induced proliferation of T_{reg} cells that precedes their involvement in immune suppression within peripheral tissues [61]. The presence of CXCR5$^+$ T_{reg} cells capable of suppressing germinal centre T helper cell-driven antibody responses suggests that the diverse subsets of effector T cells with proinflammatory function may also have subsets of T_{reg} cells with matching migration preferences [62]. Other populations of T_{reg} cells include those infiltrating the synovial tissue in RA or the airways in allergic responses [63]. The integrin αEβ7 discriminates between LN- and inflamed tissue-homing T_{reg} cells, and the latter subset was found to contain the most potent suppressors of inflammatory processes in disease models, such as antigen-induced arthritis [64]. Early studies showed that circulating and thymic human T_{reg} cells expressed CCR4 and/or CCR8 [65, 66] but, clearly, chemokine receptors do not discriminate between T_{reg} cells and pro-inflammatory T cells. The definitions of the migration patterns associated with distinct T_{reg} cells will largely depend on the discovery of reliable markers for these cells.

Migration properties in effector and memory T cells

The exact nature of immunological memory is still poorly understood. Nevertheless, cell surface markers (particularly homing-related molecules) have been extremely useful to mark and study different populations of T cells, in particular subsets relating to effector and memory T cells. It is our view that naïve T cells are a homogeneous population of T cells, that express CCR7 and L-selectin, recirculate randomly through secondary lymphoid tissues, and do not subdivide further into distinct migratory or functional subsets [1, 3]. However, antigen stimulation, proliferation (clonal expansion) and subsequent differentiation into effector or memory cells leads to subset-restricted expression of chemoattractant receptors or adhesion molecules. The cellular address code, composed of a combination of migration and adhesion molecules, fully mirrors the potential involvement of effector/memory T cells in distinct immune processes, including diverse antimicrobial responses and immune surveillance. In terms of functional criteria, the order in relatedness of the main subsets would be naïve T cells, T_{CM} cells, T_{EM} cells and effector T cells, although alternative models in the sequence of T cell differentiation have also been proposed. Still, naïve T cells share many functional and homing features with T_{CM} cells whereas effector T cells share many characteristics with T_{EM} cells.

Tissue-selectivity of memory T cells

Tissue-specific migration by T cells was first observed in the 1970s in sheep and then in mice. A hallmark finding was the discovery that antigen-experienced (memory) T cells but not naïve T cells displayed homing preferences for distinct peripheral tissues [67]. The rationale is that T cells recognising cutaneous-associated pathogens should migrate preferentially to the skin where they are likely to re-encounter their antigen, whereas T cells with selectivity for gastrointestinal pathogens would contribute to mucosal rather than cutaneous defence. The best understood examples of tissue-selective homing are related to T cell traffic in the skin and gut, but migration selectivity for other organs, such as the lung or joints, may also exist.

As discussed above, the adhesion molecules or chemokine receptors responsible for tissue migration are either absent from naïve T cells or are expressed at low levels. Moreover, distinct subsets of memory T cells are definable by adhesion molecules or chemokine receptors. For instance, newly generated skin-homing T cells express cutaneous lymphocyte-associated antigen (CLA) in combination with CCR4 and/or CCR10. By contrast, gut-homing memory T cells express high levels of the integrin $\alpha4\beta7$ in combination with CCR9 but not CCR4 and CCR10. The imprinting of gut-homing or skin-homing programs on T cells is mediated by DCs in gut-associated or skin-associated LNs, such that activation by intestinal DCs induces a "gut-tropism" [68], whereas DCs from peripheral LNs induce homing receptors in CD8 T cells that are characteristic for a "skin-tropism" [69]. Memory T cells remain responsive to alternative tissue imprinting signals, thus allowing skin- or gut-homing T cells to change their migration preferences when stimulated by DCs from alternative tissues [69]. The molecular mechanisms underlying the instalment of mutually exclusive migration profiles are not yet fully understood, although retinoic acid appears to be involved in the imprinting of gut-tropism.

Conclusions

A feature of adaptive immune responses in species such as man and mouse is the extraordinary level of sophistication, with respect to numbers of T cell subsets, their diverse functions, and their migration pathways. This is possibly why there are so many chemokines and chemokine receptors, which serve to provide the fine specificity of T cell placement in the body. A number of interesting questions have emerged. Does the migration profile of a distinct T cell indeed predetermine its function in immune processes? Is this address code, composed of a set of chemokine receptors and adhesion molecules, the basis for or the consequence of T cell differentiation? Does the remarkable combinatorial diversity in chemokine receptors reflect functional specialisation in T cells? If this is true, then the actual number of distinct T cell subsets is much larger than currently appreciated. An obvious question that follows relates to novel or less well understood grounds of T cell subset specialisation. For instance, chemokines may directly contribute to effector-to-memory T cell transition by removal of effector T cells from the site of effector T cell apoptosis. Alternatively, chemokines could effect the formation of central and peripheral tolerance by controlling localisation of T cells within distinct niches in the thymus or LNs that support T_{reg} cell differentiation. The field of T cell relocation and positioning still has numerous controversies, and many of these relate to conflicting data obtained in sheep, mice and man. Progress in this area will largely depend on the identification of orthologous cell surface markers, including receptors for chemokines. Finally, given the importance of chemokines and adhesion molecules in the control of inflammatory processes, one may propose that the next generation of anti-inflammatory drugs will target effector T cell migration. Here, it will be important to take into account species-specific differences in the immune system that prevent unfiltered extrapolation of *in vivo* findings obtained in mice to man.

References

1 Moser B, Wolf M, Walz A, Loetscher P (2004) Chemokines: multiple levels of leukocyte migration control. *Trends Immunol* 25: 75–84

2 Sallusto F, Mackay CR (2004) Chemoattractants and their receptors in homeostasis and inflammation. *Curr Opin Immunol* 16: 724–731

3 Von Andrian UH, Mackay CR (2000) T-cell function and migration. Two sides of the same coin. *N Engl J Med* 343: 1020–1034

4 Von Andrian UH, Mempel TR (2003) Homing and cellular traffic in lymph nodes. *Nat Rev Immunol* 3: 867–878

5 Butcher EC, Williams M, Youngman K, Rott L, Briskin M (1999) Lymphocyte trafficking and regional immunity. *Adv Immunol* 72: 209–253

6 Sallusto F, Lenig D, Förster R, Lipp M, Lanzavecchia A (1999) Two subsets of memory T lymphocytes with distinct homing potentials and effector functions. *Nature* 401: 708–712

7 Gunn MD, Tangemann K, Tam C, Cyster JG, Rosen SD, Williams LT (1998) A chemokine expressed in lymphoid high endothelial venules promotes the adhesion and chemotaxis of naive T lymphocytes. *Proc Natl Acad Sci USA* 95: 258–263

8 Stein JV, Rot A, Luo Y, Narasimhaswamy M, Nakano H, Gunn MD, Matsuzawa A, Quackenbush EJ, Dorf ME, Von Andrian UH (2000) The CC chemokine thymus-derived chemotactic agent 4 (TCA-4, secondary lymphoid tissue chemokine, 6Ckine, exodus-2) triggers lymphocyte function-associated antigen 1-mediated arrest of rolling T lymphocytes in peripheral lymph node high endothelial venules. *J Exp Med* 191: 61–75

9 Baekkevold ES, Yamanaka T, Palframan RT, Carlsen HS, Reinholt FP, Von Andrian UH, Brandtzaeg P, Haraldsen G (2001) The CCR7 ligand elc (CCL19) is transcytosed in high endothelial venules and mediates T cell recruitment. *J Exp Med* 193: 1105 1112

10 Muller G, Hopken UE, Stein H, Lipp M (2002) Systemic immunoregulatory and pathogenic functions of homeostatic chemokine receptors. *J Leukoc Biol* 72: 1–8

11 Banchereau J, Briere F, Caux C, Davoust J, Lebecque S, Liu YJ, Pulendran B, Palucka K (2000) Immunobiology of dendritic cells. *Annu Rev Immunol* 18: 767–811

12 Steinman RM, Hawiger D, Nussenzweig MC (2003) Tolerogenic dendritic cells. *Annu Rev Immunol* 21: 685–711

13 Schaerli P, Ebert L, Willimann K, Blaser A, Roos RS, Loetscher P, Moser B (2004) A

Skin-selective homing mechanism for human immune surveillance T cells. *J Exp Med* 199: 1265–1275
14 D'Ambrosio D, Sinigaglia F, Adorini L (2003) Special attractions for suppressor T cells. *Trends Immunol* 24: 122–126
15 Kunkel EJ, Butcher EC (2002) Chemokines and the tissue-specific migration of lymphocytes. *Immunity* 16: 1–4
16 Olaussen RW, Farstad IN, Brandtzaeg P, Rugtveit J (2001) Age-related changes in CCR9$^+$ circulating lymphocytes: are CCR9$^+$ naive T cells recent thymic emigrants? *Scand J Immunol* 54: 435–439
17 Okada T, Ngo VN, Ekland EH, Forster R, Lipp M, Littman DR, Cyster JG (2002) Chemokine requirements for B cell entry to lymph nodes and Peyer's patches. *J Exp Med* 196: 65–75
18 Phillips R, Ager A (2002) Activation of pertussis toxin-sensitive CXCL12 (SDF-1) receptors mediates transendothelial migration of T lymphocytes across lymph node high endothelial cells. *Eur J Immunol* 32: 837–847
19 Lanzavecchia A, Sallusto F (2002) Progressive differentiation and selection of the fittest in the immune response. *Nat Rev Immunol* 2: 982–987
20 Schaerli P, Loetscher P, Moser B (2001) Cutting edge: induction of follicular homing precedes effector Th cell development. *J Immunol* 167: 6082–6086
21 Ansel KM, McHeyzer-Williams LJ, Ngo VN, McHeyzer-Williams MG, Cyster JG (1999) *In vivo*-activated CD4 T cells upregulate CXC chemokine receptor 5 and reprogram their response to lymphoid chemokines. *J Exp Med* 190: 1123–1134
22 Legler DF, Loetscher M, Roos RS, Clark-Lewis I, Baggiolini M, Moser B (1998) B cell-attracting chemokine 1, a human CXC chemokine expressed in lymphoid tissues, selectively attracts B lymphocytes via BLR1/CXCR5. *J Exp Med* 187: 655–660
23 Gunn MD, Ngo VN, Ansel KM, Ekland EH, Cyster JG, Williams LT (1998) A B-cell-homing chemokine made in lymphoid follicles activates Burkitt's lymphoma receptor-1. *Nature* 391: 799–803
24 Schaerli P, Willimann K, Lang AB, Lipp M, Loetscher P, Moser B (2000) CXC chemokine receptor 5 expression defines follicular homing T cells with B cell helper function. *J Exp Med* 192: 1553–1562
25 Muller G, Hopken UE, Lipp M (2003) The impact of CCR7 and CXCR5 on lymphoid organ development and systemic immunity. *Immunol Rev* 195: 117–135
26 Förster R, Mattis AE, Kremmer E, Wolf E, Brem G, Lipp M (1996) A putative chemokine receptor, BLR1, directs B cell migration to defined lymphoid organs and specific anatomic compartments of the spleen. *Cell* 87: 1037–1047
27 Gulbranson-Judge A, MacLennan I (1996) Sequential antigen-specific growth of T cells in the T zones and follicles in response to pigeon cytochrome c. *Eur J Immunol* 26: 1830–1837
28 Garside P, Ingulli E, Merica RR, Johnson JG, Noelle RJ, Jenkins MK (1998) Visualization of specific B and T lymphocyte interactions in the lymph node. *Science* 281: 96–99
29 Breitfeld D, Ohl L, Kremmer E, Ellwart J, Sallusto F, Lipp M, Forster R (2000) Follicu-

lar B helper T cells express CXC chemokine receptor 5, localize to B cell follicles, and support immunoglobulin production. *J Exp Med* 192: 1545–1552
30 Langenkamp A, Nagata K, Murphy K, Wu L, Lanzavecchia A, Sallusto F (2003) Kinetics and expression patterns of chemokine receptors in human CD4+ T lymphocytes primed by myeloid or plasmacytoid dendritic cells. *Eur J Immunol* 33: 474–482
31 Ebert LM, Horn MP, Lang AB, Moser B (2004) B cells alter the phenotype and function of follicular-homing CXCR5+ T cells. *Eur J Immunol* 34: 3562–3571
32 Mak TW, Shahinian A, Yoshinaga SK, Wakeham A, Boucher LM, Pintilie M, Duncan G, Gajewska BU, Gronski M, Eriksson U et al (2003) Costimulation through the inducible costimulator ligand is essential for both T helper and B cell functions in T cell-dependent B cell responses. *Nat Immunol* 4: 765–772
33 Mackay CR (2001) Chemokines: immunology's high impact factors. *Nat Immunol* 2: 95–101
34 Butcher EC, Picker LJ (1996) Lymphocyte homing and homeostasis. Science 272: 60–66
35 Sallusto F, Lanzavecchia A, Mackay CR (1998) Chemokines and chemokine receptors in T-cell priming and Th1/Th2-mediated responses. *Immunol Today* 19: 568–574
36 Syrbe U, Siveke J, Hamann A (1999) Th1/Th2 subsets: distinct differences in homing and chemokine receptor expression? *Springer Semin Immunopathol* 21: 263–285
37 Sallusto F, Lenig D, Mackay CR, Lanzavecchia A (1998) Flexible programs of chemokine receptor expression on human polarized T helper 1 and 2 lymphocytes. *J Exp Med* 187: 875–883
38 Bonecchi R, Bianchi G, Bordignon PP, D'Ambrosio D, Lang R, Borsatti A, Sozzani S, Allavena P, Gray PA, Mantovani A et al (1998) Differential expression of chemokine receptors and chemotactic responsiveness of type 1 T helper cells (Th1s) and Th2s. *J Exp Med* 187: 129–134
39 Loetscher P, Uguccioni M, Bordoli L, Baggiolini M, Moser B, Chizzolini C, Dayer J-M (1998) CCR5 is characteristic of Th1 lymphocytes. *Nature* 391: 344–345
40 Qin SX, Rottman JB, Myers P, Kassam N, Weinblatt M, Loetscher M, Koch AE, Moser B, Mackay CR (1998) The chemokine receptors CXCR3 and CCR5 mark subsets of T cells associated with certain inflammatory reactions. *J Clin Invest* 101: 746–754
41 Paxton WA, Kang S (1998) Chemokine receptor allelic polymorphisms: relationships to HIV resistance and disease progression. *Semin Immunol* 10: 187–194
42 Gomez-Reino JJ, Pablos JL, Carreira PE, Santiago B, Serrano L, Vicario JL, Balsa A, Figueroa M, De Juan MD (1999) Association of rheumatoid arthritis with a functional chemokine receptor, CCR5. *Arthritis Rheum* 42: 989–992
43 Austrup F, Vestweber D, Borges E, Löhning M, Bräuer R, Herz U, Renz H, Hallmann R, Scheffold A, Radbruch A et al (1997) P- and E-selectin mediate recruitment of T-helper-1 but not T-helper-2 cells into inflamed tissues. *Nature* 385: 81–83
44 Borges E, Tietz W, Steegmaier M, Moll T, Hallmann R, Hamann A, Vestweber D (1997) P-selectin glycoprotein ligand-1 (PSGL-1) on T helper 1 but not on T helper 2 cells binds to P-selectin and supports migration into inflamed skin. *J Exp Med* 185: 573–578
45 Xie HJ, Lim YC, Luscinskas FW, Lichtman AH (1999) Acquisition of selectin binding

and peripheral homing properties by CD4+ and CD8+ T cells. *J Exp Med* 189: 1765–1775

46 Maly P, Thall AD, Petryniak B, Rogers GE, Smith PL, Marks RM, Kelly RJ, Gersten KM, Cheng GY, Saunders TL et al (1996) The a(1,3)Fucosyltransferase Fuc-TVII controls leukocyte trafficking through an essential role in L-, E-, and P-selectin ligand biosynthesis. *Cell* 86: 643–653

47 Lim YC, Henault L, Wagers AJ, Kansas GS, Luscinskas FW, Lichtman AH (1999) Expression of functional selectin ligands on Th cells is differentially regulated by IL-12 and IL-4. *J Immunol* 162: 3193–3201

48 Wagers AJ, Waters CM, Stoolman LM, Kansas GS (1998) Interleukin 12 and interleukin 4 control T cell adhesion to endothelial selectins through opposite effects on alpha1, 3-fucosyltransferase VII gene expression. *J Exp Med* 188: 2225–2231

49 Sallusto F, Mackay CR, Lanzavecchia A (1997) Selective expression of the eotaxin receptor CCR3 by human T helper 2 cells. *Science* 277: 2005–2007

50 Gerber BO, Zanni MP, Uguccioni M, Loetscher M, Mackay CR, Pichler WJ, Yawalkar N, Baggiolini M, Moser B (1997) Functional expression of the eotaxin receptor CCR3 in T lymphocytes co-localizing with eosinophils. *Curr Biol* 7: 836–843

51 Sundrud MS, Grill SM, Ni D, Nagata K, Alkan SS, Subramaniam A, Unutmaz D (2003) Genetic reprogramming of primary human T cells reveals functional plasticity in Th cell differentiation. *J Immunol* 171: 3542–3549

52 Jose PJ, Griffiths-Johnson DA, Collins PD, Walsh DT, Moqbel R, Totty NF, Truong O, Hsuan JJ, Williams TJ (1994) Eotaxin: A potent eosinophil chemoattractant cytokine detected in a guinea pig model of allergic airways inflammation. *J Exp Med* 179: 881–887

53 Ponath PD, Qin SX, Ringler DJ, Clark-Lewis I, Wang J, Kassam N, Smith H, Shi XJ, Gonzalo JA, Newman W et al (1996) Cloning of the human eosinophil chemoattractant, eotaxin – Expression, receptor binding, and functional properties suggest a mechanism for the selective recruitment of eosinophils. *J Clin Invest* 97: 604–612

54 Chensue SW, Lukacs NW, Yang TY, Shang X, Frait KA, Kunkel SL, Kung T, Wiekowski MT, Hedrick JA, Cook DN et al (2001) Aberrant *in vivo* T helper type 2 cell response and impaired eosinophil recruitment in CC chemokine receptor 8 knockout mice. *J Exp Med* 193: 573–584

55 Goya I, Villares R, Zaballos A, Gutierrez J, Kremer L, Gonzalo JA, Varona R, Carramolino L, Serrano A, Pallares P et al (2003) Absence of CCR8 does not impair the response to ovalbumin-induced allergic airway disease. *J Immunol* 170: 2138–2146

56 Chung CD, Kuo F, Kumer J, Motani AS, Lawrence CE, Henderson WR Jr, Venkataraman C (2003) CCR8 is not essential for the development of inflammation in a mouse model of allergic airway disease. *J Immunol* 170: 581–587

57 Chong BF, Murphy JE, Kupper TS, Fuhlbrigge RC (2004) E-selectin, thymus- and activation-regulated chemokine/CCL17, and intercellular adhesion molecule-1 are constitutively coexpressed in dermal microvessels: a foundation for a cutaneous immunosurveillance system. *J Immunol* 172: 1575–1581

57a Reiss Y, Proudfoot AE, Power CA, Campbell JJ, Butcher EC (2001) CC chemokine receptor (CCR)4 and the CCR10 ligand cutaneous T cell-attracting chemokine (CTACK) in lymphocyte trafficking to inflamed skin. *J Exp Med* 194: 1541–1547
58 Biedermann T, Schwarzler C, Lametschwandtner G, Thoma G, Carballido-Perrig N, Kund J, De Vries JE, Rot A, Carballido JM (2002) Targeting CLA/E-selectin interactions prevents CCR4-mediated recruitment of human Th2 memory cells to human skin *in vivo*. *Eur J Immunol* 32: 3171–3180
59 Sakaguchi S (2004) Naturally arising $CD4^+$ regulatory t cells for immunologic self-tolerance and negative control of immune responses. *Annu Rev Immunol* 22: 531–562
60 Stock P, Akbari O, Berry G, Freeman GJ, DeKruyff RH, Umetsu DT (2004) Induction of T helper type 1-like regulatory cells that express Foxp3 and protect against airway hyper-reactivity. *Nat Immunol* 5: 1149–1156
60a Szanya V, Ermann J, Taylor C, Holness C, Fathman CG (2002) The subpopulation of CD4+CD25+ splenocytes that delays adoptive transfer of diabetes expresses L-selectin and high levels of CCR7. *J Immunol* 169: 2461–2465
61 Apostolou I, Sarukhan A, Klein L, Von Boehmer H (2002) Origin of regulatory T cells with known specificity for antigen. *Nat Immunol* 3: 756–763
62 Lim HW, Hillsamer P, Kim CH (2004) Regulatory T cells can migrate to follicles upon T cell activation and suppress GC-Th cells and GC-Th cell-driven B cell responses. *J Clin Invest* 114: 1640–1649
63 Cao D, van Vollenhoven R, Klareskog L, Trollmo C, Malmstrom V (2004) $CD25^{bright}CD4^+$ regulatory T cells are enriched in inflamed joints of patients with chronic rheumatic disease. *Arthritis Res Ther* 6: R335–R346
64 Huehn J, Siegmund K, Lehmann JC, Siewert C, Haubold U, Feuerer M, Debes GF, Lauber J, Frey O, Przybylski GK et al (2004) Developmental stage, phenotype, and migration distinguish naive- and effector/memory-like $CD4^+$ regulatory T cells. *J Exp Med* 199: 303–313
65 Iellem A, Mariani M, Lang R, Recalde H, Panina-Bordignon P, Sinigaglia F, D'Ambrosio D (2001) Unique chemotactic response profile and specific expression of chemokine receptors CCR4 and CCR8 by CD4(+)CD25(+) regulatory T cells. *J Exp Med* 194: 847–853
66 Annunziato F, Cosmi L, Liotta F, Lazzeri E, Manetti R, Vanini V, Romagnani P, Maggi E, Romagnani S (2002) Phenotype, localization, and mechanism of suppression of CD4(+)CD25(+) human thymocytes. *J Exp Med* 196: 379–387
67 Mackay CR (1991) T-cell memory: the connection between function, phenotype and migration pathways. *Immunol Today* 12: 189–192
68 Mora JR, Bono MR, Manjunath N, Weninger W, Cavanagh LL, Rosemblatt M, Von Andrian UH (2003) Selective imprinting of gut-homing T cells by Peyer's patch dendritic cells. *Nature* 424: 88–93
69 Mora JR, Cheng G, Picarella D, Briskin M, Buchanan N, Von Andrian UH (2005) Reciprocal and dynamic control of CD8 T cell homing by dendritic cells from skin- and gut-associated lymphoid tissues. *J Exp Med* 201: 303–316

Lymphocyte homing to peripheral epithelial tissues

William W. Agace[1] *and Bernhard Homey*[2]

[1]Immunology Section, Lund University, BMC I-13, 22184 Lund, Sweden; [2]Department of Dermatology, Heinrich-Heine-University, Duesseldorf, Germany

Introduction

Epithelial tissues represent the interface between the environment and the host. They are subject to continuous insults that include mechanical injury, ultraviolet (UV) irradiation, chemicals and microbes. The integrity of the host critically depends on the adequate protection against these hazardous events. During evolution epithelial tissues developed specialised immunological structures such as mucosa-associated lymphoid tissues (MALT) or skin-associated lymphoid tissues (SALT) which together with patrolling leukocyte subsets work as sentinels at the inner and outer surface of the human body. Among patrolling leukocytes, effector memory T cells take a centre stage and show tissue-specific migration patterns. To date, at least two distinct populations of effector memory T cells have been identified. Memory T cells expressing $\alpha_4\beta_7$ integrins preferentially migrate into the gut while the cutaneous lymphocyte associated antigen (CLA) identifies a subset of skin-homing memory T cells. Here, we provide an overview of current concepts how chemokines regulate lymphocyte trafficking into distinct epithelial tissues.

The role of chemokines in lymphocyte localisation to the gut mucosa

The intestinal surface is comprised of a single layered epithelium that separates the contents of the lumen from the intestinal lamina propria (LP) (Fig. 1). The inductive sites of the small intestine are Peyers Patches (PP) or isolated lymphoid follicles (ILF) which lie directly underneath this epithelium, and mesenteric lymph nodes (MLN) whose afferent lymphatics drain the intestinal LP (Fig. 1). Antigen enters PP via specialised microfold (M) cells within the Follicular Associated Epithelium (FAE) where it is taken up and processed by dendritic cells in the sub-epithelial dome (SED) for presentation to T cells. In contrast, luminal antigen accessing the intestinal LP is transported to MLN via draining lymphatics either directly or with-

Chemokine Biology – Basic Research and Clinical Application, Volume I
edited by Bernhard Moser, Gordon L. Letts and Kuldeep Neote

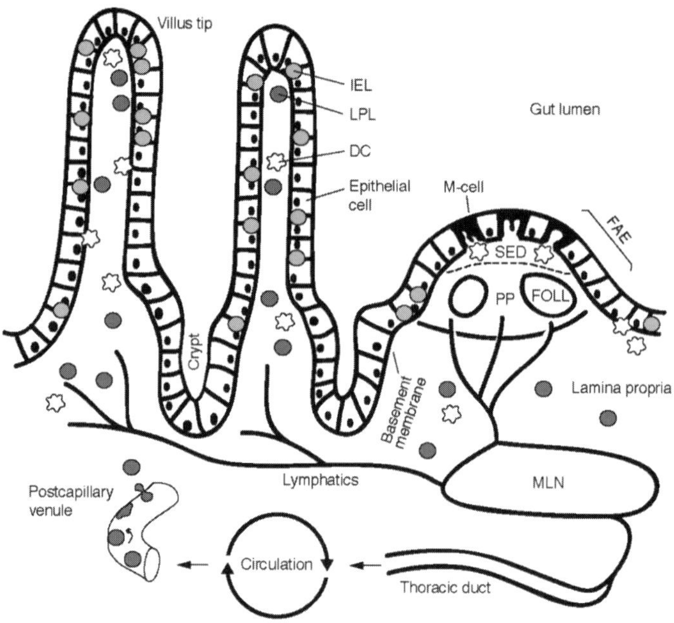

Figure 1
Schematic overview of the inductive and effector sites within the intestinal mucosa
DC, dendritic cell; FAE, follicle associated epithelium; IEL, FOLL, Follicle; intraepithelial lymphocyte; LPL, lamina propria lymphocyte; PP, peyers patch; MLN, mesenteric lymph node; SED, subepithelial dome. Arrows indicate the circulation route of GALT primed T cells.

in mobilised DCs. Presentation of luminal antigen in the context of co-stimulation, to naïve T cells within PP or MLN, induces their activation and proliferation and a proportion of these cells re-enter the circulation via the thoracic duct and localise to the two effector sites of the intestinal mucosa, the intestinal epithelium (as intraepithelial lymphocytes, IEL) and lamina propria (as lamina propria lymphocytes, LPL). In the mouse, T lymphocyte entry to the intestinal mucosa is largely dependent on the integrin $\alpha_4\beta_7$ through its interactions with MadCAM-1 on intestinal microvascular endothelial cells [1–3]. Furthermore, activated/memory $\alpha_4\beta_7^+$ T cells in human peripheral blood preferentially contain memory for intestinal antigens [4].

The intestinal epithelium and lamina propria contains a large number of previously activated/memory T cells under steady state conditions, presumably as a result of the high antigenic load within the intestinal lumen. LPL consist mainly of $CD4^+$ T cells and IgA secreting plasma cells that are thought to enter the intestinal mucosa following their activation in gut associated lymphoid tissue (GALT). IEL are primarily $CD8^+$ T cells and include both conventional major histocompatibility complex (MHC) class I restricted $CD8\alpha\beta^+$ $TCR\alpha\beta^+$ cells at least some of which enter this site following priming in GALT and unconventional $CD8\alpha\alpha^+$ $TCR\alpha\beta^+$ and $CD8\alpha\alpha^+$ $TCR\gamma\delta^+$ IEL that are unique to the intestine [5, 6]. These latter populations appear to recognise non-classical MHC class Ib molecules and their ontogeny and state of differentiation at time of entry into the epithelium remains a subject of debate.

Chemokine receptor expression on T cells resident in intestinal effector sites

Human small intestinal LP T cells and IEL express a restricted array of chemokine receptors including CXCR3, CXCR4, CCR5, and CCR9 but not CXCR1, CXCR2, CCR1, CCR3, 4, 7, 8 and 10 [7–12]. CCR6 is expressed on a subset of murine and human small intestinal LPL ([13]; Stenstad et al., manuscript in preparation) but on few if any IEL [13, 14]. CXCR6 is expressed by a large proportion of murine IEL, although its expression on LPL and human intestinal lymphocytes has yet to be determined [15, 16]. In addition a variable number of $CD4^+$LPL express the chemokine receptor CCR2 [7, 17]. Human colonic lymphocytes express CXCR3 and CCR5 but not CCR4, and a small proportion express CCR2 [17, 18]. Remarkably CCR9 is largely absent from colonic T cells and T cells isolated from other peripheral tissues, including the skin [19–21]. Combined, these studies suggest that chemokine receptors may contribute to intestinal T cell localisation and/or function within distinct segments of the intestine (CCR9 for the small intestine), within the LP *versus* the epithelium (CCR2 and CCR6), or have more global functions within intestinal effector sites (CXCR3 and CCR5).

CCR9/CCL25 mediates T cell recruitment to the small intestine

The selective expression of CCR9 on previously activated/memory $\alpha_4\beta_7^+$ 'gut homing' T cells in peripheral blood and on small intestinal lymphocytes [19–21], and the constitutive and selective expression of its ligand, CCL25, by small intestinal epithelial cells [19, 22] lead to the suggestion that this chemokine receptor/chemokine pair plays a unique role in small intestinal immune responses. Consistent with this suggestion, $CCR9^{-/-}$ mice have reduced numbers of small intestinal IEL, primarily among the $CD8\alpha\alpha^+TCR\gamma\delta^+$ IEL subset [23, 24], although the total number of $CD8\alpha\beta^+$ $TCR\alpha\beta^+$ IEL and $CD4^+$ LPL appears normal [23–25]. Studies of young mice treated with neutralising anti-CCL25 antibody [26], or CCL25 intrakine mice, whose T cells fail to express CCR9 [27], have demonstrated an important role for CCL25/CCR9 in the generation of both the CD8$\alpha\alpha^{+}$TCR$\alpha\beta^{+}$ and CD8$\alpha\alpha^{+}$TCR$\gamma\delta^{+}$ IEL compartment although how CCR9 regulates the generation of this compartment remains to be determined.

CCR9/CCL25 is also important for the initial recruitment of conventional effector CD8$\alpha\beta^{+}$ T cells to the small intestinal epithelium following their priming in GALT. Thus anti-CCL25 antibody reduced effector CD8$\alpha\beta^{+}$ T cell localisation to the intestinal epithelium and in a competitive TCR transgenic adoptive transfer model CCR9$^{-/-}$ effector CD8$\alpha\beta^{+}$ T cells were selectively disadvantaged in their ability to localise to this site [21, 28]. While CCL25 mRNA is expressed primarily by small intestinal epithelial cells, CCL25 protein has been detected on small intestinal microvascular endothelium by immunohistochemistry [20, 29], indicating that CCL25 may function in part by mediating CCR9^{+} T cell arrest on lamina propria vessels [30]. Additionally CCR9 appears to regulate the expression and function of the mucosal integrin $\alpha_E\beta_7$ on small intestinal IEL [14].

Importantly, no equivalent system has been described to mediate effector T cell recruitment to the colon, and the role of chemokines in effector T lymphocyte localisation to this site remains unclear.

Role of additional chemokine receptors in homeostatic intestinal T cell localisation and function

CXCR3 and CCR5 ligands in general appear to be expressed at low levels in healthy small intestine and colon (although expression of some ligands such as CXCL11 have yet to be examined) [31–35]. Thus it seems unlikely that CCR5 and CXCR3 contribute in a major way to intestinal lymphocyte localisation/function under steady state conditions. Indeed since these receptors are expressed by lymphocytes in many non-lymphoid tissue [18] their expression on intestinal lymphocytes may simply reflect the activation status of these cells.

The CCR6 ligand, CCL20, is constitutively expressed by intestinal epithelial cells, particularly by the FAE [36], and CCR6$^{-/-}$ mice display increased numbers of LPL, IEL, particularly of the CD8$\alpha\alpha^{+}$TCR$\alpha\beta^{+}$ subset, and reduced PP size [37, 38]. Thus CCR6/CCL20 plays a critical role in maintaining intestinal T cell homeostasis. Since mature IEL fail to express CCR6 [13, 14], CCR6 is unlikely to regulate mature IEL localisation and function. However CCR6 was recently implicated in regulating the generation of non-conventional IEL within intestinal cryptopatches [13], putative sites of extrathymic T cell development, although how CCR6 functions in this process remains unclear. Since, CCL20 is also expressed by DCs, DC derived CCL20 may regulate CCR6^{+} CD4^{+} T cell interactions with antigen presenting cells in the LP. Finally since epithelial derived CCL20 has also been proposed to regulate dendritic cell (DC) influx into the intestinal epithelium [37], although its involvement in regulating DC migration into the FAE has recently been questioned [39], CCR6/CCL20 may act indirectly via DCs in the control of intestinal T cell numbers.

The CCR2 ligands CCL2, CCL7 and CCL8 are expressed in the healthy human intestine, however results regarding their levels of expression and cellular source vary between studies [34, 35, 40–43]. It seems likely that CCR2 ligands influence the localisation and function of the few $CCR2^{+}$ $CD4^{+}$ T cells in the lamina propria. The numbers and populations of IEL and LPL in $CCR2^{-/-}$ mice have not been reported; however, CCR2 and its ligand CCL2 have been implicated in high dose oral tolerance although they appear, in this case, to be functioning at the level of antigen presentation in gut inductive sites [44].

The CXCR4 ligand, CXCL12, is constitutively expressed by intestinal epithelial cells, intestinal microvascular endothelial vessels, and pericytes surrounding these vessels [9, 45] and can induce $\alpha_4\beta_7$ integrin mediated T cell adhesion to MadCAM-1 and Fibronectin [46]. Together these reports suggest a potential role for CXCR4/CXCL12 in mediating T cell recruitment to the intestinal mucosa. Since CXCR4 is expressed on a wide variety of T cells it is unlikely to contribute to the selective recruitment of effector T cell subsets to the intestine, however such selectivity could be provided by the integrin $\alpha_4\beta_7$.

Recruitment of IgA immunoblasts to intestinal effector sites

IgA secreting plasma cells in the LP derive from circulating IgA immunoblasts that have been generated in intestinal inductive sites. Two chemokine receptors, CCR9 and CCR10, have been implicated in the recruitment of IgA immunoblasts to the intestinal mucosa. CCR9 is expressed on a subset of human circulating IgA immunoblasts and IgA plasma cells in the small intestine [10]. In the mouse, IgA immunoblasts from the spleen and MLN as well as $B220^{int}IgA^{+}$ cells in the small intestinal LP, respond to CCL25 while terminally differentiated $B220^{-}IgA^{+}$ small intestinal plasmablasts failed to migrate to this chemokine [47]. Importantly, IgA^{+} plasma cells are reduced in the small intestine but not colon of $CCR9^{-/-}$ mice [25] and anti-CCL25 antibody significantly inhibited CT specific IgA immunoblast localisation to the small intestine but not the colon [29]. Thus CCR9 appears to play a role in the selective localisation of IgA^{+} immunoblasts to the small intestinal mucosa.

CCR10 is expressed on human circulating IgA immunoblasts and virtually all IgA plasma cells in the small and large intestine, appendix, tonsil and salivary gland [10] while its ligand CCL28, is constitutively expressed by epithelial cells at these sites [48, 49]. Furthermore anti-CCL28 antibody inhibited the localisation of IgA plasmablasts to the murine small intestine and colon [29]. Thus CCR10/CCL28 appears to play a more global role in recruiting IgA immunoblasts to mucosal surfaces.

Lymphocyte recruitment to the inflamed intestinal mucosa

Intestinal T cell numbers increase dramatically in the setting of intestinal inflammation such as inflammatory bowel disease (IBD, Crohn's disease and ulcerative colitis) and enteropathies associated with food hypersensitivity such as Coeliac's disease, and are thought to contribute in a primary way to disease pathogenesis. While ulcerative colitis is restricted to the colon, Crohn's disease can develop throughout the intestine, primarily in the distal ileum and ascending colon. A similar and wide range of chemokines are induced in ulcerative colitis and Crohn's disease [40] including the T cell chemoattractants CCL2 [34, 35, 40, 42, 43], CCL3 [33, 40], CCL4 [33, 40], CCL5 [33, 35, 50], CCL7 [34, 40, 41], CCL8 [40], CCL20 [51, 52], CXCL9 [32] and CXCL10 [32–34]. A notable exception is CCL25 that is expressed in small bowel Crohn's but not colonic Crohn's disease or Ulcerative Colitis [53]. The proportion of LP and MLN T cells that express CCR9 is significantly reduced in small bowel Crohn's compared to healthy intestine [53]. While this reduction may result from increased activation induced cell death of $CCR9^+$ T cells as originally proposed [53], an equally plausible explanation is that alternative chemokine receptors play a more dominant role in recruiting effector T cells to the inflamed small intestine. Studies from knockout mice or antibody neutralisation experiments have implicated a role for several chemokine receptors in regulating disease severity in animal models of IBD, including CCR2, CCR5, CCR6, and CXCR3 [54–58]. However whether these receptors regulate T cell localisation to and function within the inflamed intestine remains unclear. Because of the wide range of chemokines induced in the intestine during inflammation, it seems likely that there is some redundancy in chemokine receptors usage regarding T cell localisation and function within the inflamed intestine and that the importance of a given receptor will vary depending on the local inflammatory conditions. In this regard, CCR5 and its ligands CCL3 and CCL4, have been implicated in the recruitment of *Toxoplasma gondii* primed $CD8\alpha\beta^+$ T cells to the inflamed small intestinal epithelium [59], however in a dextran sulfate sodium (DSS) induced colitis model $CD4^+$ T cell numbers actually increase in the intestinal mucosa of $CCR5^{-/-}$ compared to WT mice [58]. Finally, while relatively few studies have directly compared chemokine receptor expression on T cells isolated from healthy and inflamed intestine, $CCR2^+CD4^+$ LP T cells were recently shown to increase in number in ileal but not colonic Crohn's disease or Ulcerative Colitis [17], implicating a selective role for CCR2 in the recruitment and/or function of $CD4^+$ cells during small bowel Crohn's.

The role of chemokines in lymphocyte localisation to the skin

The skin can be divided into an avascular epidermis and a collagen-rich, vessel-containing dermal compartment (Fig. 2). Effector memory T cells traffic between sec-

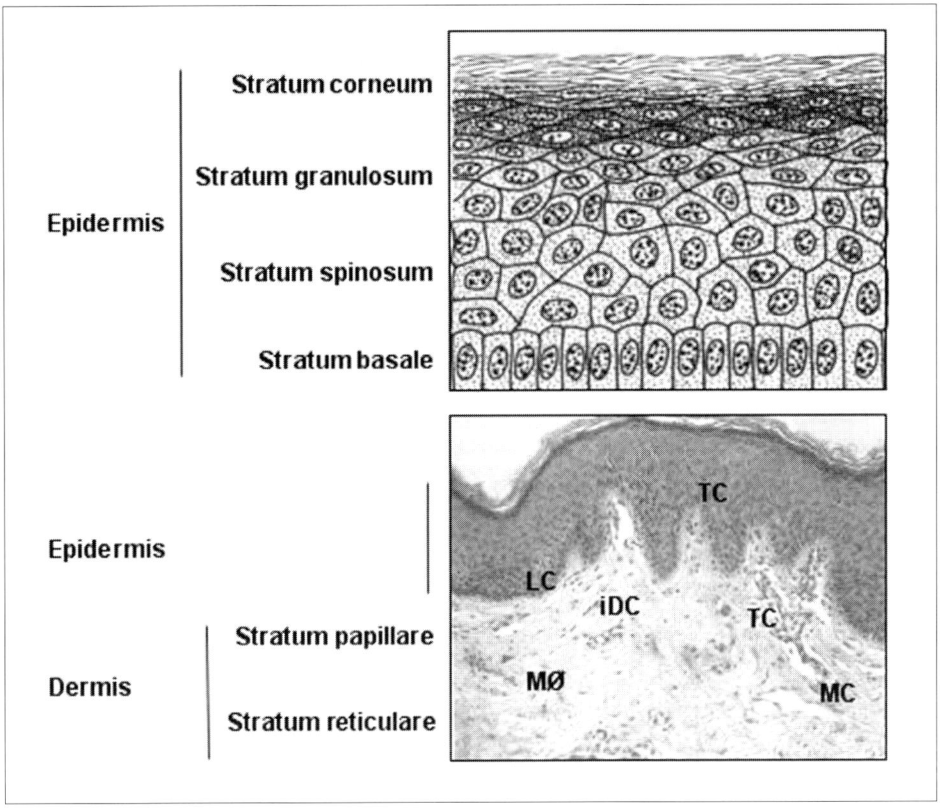

Figure 2
Schematic overview of skin
The skin can be divided into an epidermal and dermal compartment. Keratinocytes within the epidermis divide within the stratum basale and differentiate through the stratum spinosum and granulosum to finally form an acellular, keratin-rich stratum corneum. The dermis is divided in a vessel-rich stratum papillare and a matrix/collagen-rich stratum reticulare. Under homeostatic conditions Langerhans cells (LC) and few T lymphocytes (TC) are found within the epidermis. Within the dermal compartment interstitial dendritic cells (iDC) and macrophages (MØ) reside together with mast cells (MC) and patrolling T lymphocytes (TC). During inflammation the composition of leukocytes within the skin changes in quantity and quality. Elevated numbers of LC and TC are found within the epidermis and the frequency of iDC, MØ, MC, plasmacytoid dendritic cells and eosinophils may increase within the dermis.

ondary lymphoid organs and the skin. Within dermal microvessels, they interact with endothelial cells, perform transendothelial migration and enter perivascular pockets. From perivascular spaces, sustained matrix-bound gradients of chemoattractive proteins direct lymphocytes into subepidermal or intraepidermal locations. In humans, the cutaneous lymphocyte associated antigen (CLA) characterises a subset of skin-homing memory T cells. 80–90% of memory T cells in inflammatory skin lesions express CLA. In contrast, only 10–15% of the pool of circulating T cells are CLA positive. CLA^+ T lymphocytes never exceed 5% of lymphocytes within noncutaneous inflamed sites [60–62]. These observations suggest that an active and specific recruiting process focused on CLA^+ memory T cells is present in inflammatory skin lesions. Furthermore, Santamaria and co-workers showed that specific responses to common skin-associated allergens, including nickel and house dust mite, are restricted to CLA^+ T cells [63]. CLA interacts with E-selectin and mediates the rolling of distinct leukocyte subsets along the vascular endothelium. E-selectin is not skin-specific but is expressed on inflamed endothelium of various tissues. Hence, other skin-specific factors must regulate the tissue-specific homing capacity of CLA^+ memory T cells.

Chemokine receptor expression on circulating CLA^+ skin-homing memory T cells

Skin-homing memory T cells are equipped with a large panel of chemokine receptors including CCR3, CCR4, CCR5, CCR6, CCR7, CCR8, CCR10, CXCR3, and CXCR4 [11, 18, 64–68]. In particular, CCR4 and CCR10 show preferential expression on the surface of circulating CLA^+ skin-homing T cells [18, 64, 67]. While the majority of circulating $CD4^+CLA^+$ memory T cells express CCR4, only a subset (30–40%) of CLA^{high} memory T cells are $CCR10^+$ [64, 67]. These $CLA^{high}CCR10^+$ T cells of healthy individuals can act as both "central" and "effector" memory T cells, have access to both secondary lymphoid organs and the skin compartment and secrete TNF-α and IFN-γ upon *in vitro* stimulation [67].

Lymphocyte recruitment during skin homeostasis

A variety of chemokines including CCL1, CCL20, CCL27, CXCL12, and CXCL14 show homeostatic expression in healthy human skin [11, 64, 69, 70]. With regard to the recruitment of lymphocytes two chemokine/chemokine receptor pairs, CCL27/CCR10 and CCL1/CCR8 are of particular interest.

Recent studies have identified the novel skin-specific CC chemokine CCL27, which is exclusively produced by epidermal keratinocytes [64, 70]. Under homeostatic conditions, basal keratinocytes abundantly express CCL27 protein which is subsequently secreted into dermal compartments [64]. CCL27 binds the formerly orphan G-protein coupled receptor GPR-2 which has been renamed CCR10 [71]. *In vitro*, CCL27 preferentially attracts $CD4^+CLA^+$ memory T cells [70]. CCL27 shows a high binding affinity to extracellular matrix proteins and is displayed on cutaneous vascular endothelium, a phenomenon which is explained by the observation that chemokines are transported across endothelium to participate in leukocyte arrest [64, 72, 73]. Moreover, chemokines presented by endothelial cell-associated proteoglycans mediate firm adhesion as well as transendothelial migration. Previous observations suggest that binding to extracellular matrix prolongs the half-live of chemokines and increases their biological activity. Recent observations indicate that CCL27 immobilises on extracellular matrix and the surface of dermal endothelial cells and mediates the adhesion of lymphocytes [64]. Hence, endothelial cell-bound CCL27 may mediate firm adhesion and initiate transendothelial migration, while CCL27 on dermal extracellular matrix and fibroblasts may sustain a chemokine gradient directing skin-infiltrating lymphocytes from perivascular pockets to subepidermal and intraepidermal locations.

Recently, Schaerli et al. showed that the majority of human T cells in healthy skin express the chemokine receptor CCR8 and respond to its specific ligand CCL1 [11]. Normal human skin-derived $CD4^+$ and $CD8^+$ T cell subsets expressed CCR8 but $CD8^+$ lymphocytes displayed higher CCR8 surface expression and increased chemotactic responsiveness towards CCL1 gradients [11]. The majority of skin-derived $CCR8^+$ T cells expressed CLA but lacked expression of CCR4 and CCR7. These $CCR8^+$ T cells were absent in small intestine and colon tissues and represented only a very small population (<2%) in the peripheral blood. Cutaneous $CCR8^+$ T cells co-expressed CD45RO and CD45RA, displayed a pre-activated phenotype (CD69) and secreted cytokines such as TNF-α and IFN-γ but lacked markers of cytolytic T cells. Secretion of IL-4, IL-10 and TGF-β was low to undetectable, arguing against a strict association of CCR8 with either Th2 or regulatory T cell subsets. Importantly, the specific ligand for CCR8, CCL1, is constitutively expressed at strategic cutaneous locations, including dermal microvessels and epidermal antigen presenting cells. In summary, the interaction of CCL1 and CCR8 may contribute to the immune surveillance of the skin in multiple ways [11]. Endothelial cell-derived CCL1 may support the steady-state extravasation of circulating $CCR8^+$ precursors. Subsequently, Langerhans cell-derived CCL1 may direct $CCR8^+$ T cells from perivascular spaces into the epidermis, ensuring the encounter of immune surveillance T cells with epidermal antigen presenting cells.

Hence, CCL27/CCR10 and CCL1/CCR8 may represent complementary systems that support the recruitment of $CD4^+$ or $CD8^+$ memory T lymphocytes to cutaneous sites under homeostatic conditions.

Lymphocyte recruitment to inflamed skin

Accumulating evidence indicates that skin-infiltrating T cells play a pivotal role during the initiation and maintenance of inflammatory and autoimmune skin diseases, such as psoriasis or atopic dermatitis [74–77]. Hence, the understanding of mechanisms mediating memory T cell recruitment to the skin may identify promising targets for the development of novel therapeutics.

Atopic dermatitis

Atopic dermatitis is a chronic or chronically relapsing inflammatory skin disease with eczematous lesions demonstrating typical morphology and distribution, severe pruritus, elevated serum IgE, the presence of allergen-specific IgE, and peripheral blood eosinophilia [74]. The prevalence of atopic dermatitis rapidly increased during the past decades and is currently ranging between 10–20% in children and 1–3% in adults. Histopathologically, the lesional skin of atopic dermatitis patients shows a dermal infiltrate consisting of mainly activated CLA^+ memory T cells (CD4 > CD8) and antigen-presenting cells (APC) [74]. Among the APC population, lesional skin shows increased numbers of Langerhans cells (LC), inflammatory dendritic epidermal cells (IDEC), as well as dermal DCs which show markedly up-regulated expression of Fc receptors for IgE on their cell surface [74]. Moreover, dermal sites of atopic skin show extensive deposition of eosinophil-derived proteins or more rarely intact eosinophils [74]. Exposure to allergens, e.g., house dust mite antigens, or microbial products plays an important role in the initiation and maintenance of atopic skin inflammation. In early phases of the disease, memory T cells with a Th2 phenotype infiltrate the atopic skin, however, chronic lichenified atopic dermatitis lesions are characterised by the dominance of skin-infiltrating Th1 cells [74, 78].

In the past decade, numerous studies identified chemokines associated with atopic dermatitis (Tab. 1). These chemokines include CCL2, CCL3, CCL4, CCL5, CCL11, CCL13, CCL17, CCL18, CCL20, CCL22, CCL26, CCL27 and CX3CL1. Notably, serum levels of CCL11, CCL17, CCL18, CCL22, CCL26, CCL27 and CX3CL1 directly correlated with disease severity suggesting an important role in the immunopathogenesis of atopic dermatitis. Among these chemokines, CCL17, CCL18, CCL22 and CCL27 are likely candidates to critically regulate the recruitment of memory T cells to sites of atopic skin inflammation (Tab. 1). Patients suffering from atopic dermatitis show increased CCL27 protein production within the epidermis and the vast majority of skin-infiltrating lymphocytes (>90%) express CCR10 [64]. *In vivo*, intracutaneous injection of CCL27 induced the accumulation of $CD3^+$ T cells [64]. Conversely, neutralisation of CCL27-CCR10 interactions impaired memory T cell recruitment to the skin and suppressed allergen-specific skin inflammation in mouse models mimicking allergic contact dermatitis and atopic dermatitis [64].

Th2 cell lines and clones isolated from lesional skin of atopic dermatitis patients abundantly express CCR4 and show little or no CCR3, CCR8, and CXCR3 on their cell surface [79–81]. To date, there are two known ligands for CCR4, CCL17 and

Table 1 - Chemokines associated with lymphocyte recruitment to healthy and inflamed skin

Chemokine		Origin	References
Healthy skin			
CCL1	I-309	Dendritic cells, endothelial cells	[11]
CCL27	CCL27	Keratinocytes	[64, 70, 71]
Atopic dermatitis			
CCL17	TARC	Endothelial cells	[83, 85]
CCL18	PARC	Dendritic cells, keratinocytes	[90, 92]
CCL22	MDC	Macrophages, dendritic cells	[107]
CCL27	CTACK	Keratinocytes	[64, 108]
Psoriasis			
CCL17	TARC	Skin, endothelial cells	[80, 96]
CCL20	MIP-3α	Keratinocytes, fibroblasts, endothelial cells, dendritic cells	[66]
CCL27	CTACK	Keratinocytes	[64, 108]
CXCL9	Mig	Skin, keratinocytes	[96, 109]
CXCL10	IP-10	Keratinocytes, endothelial cells	[96, 109]

CCL22 [82]. CCL17 and CCL22 are produced by different cell types. In humans, the major source of CCL17 in the skin are dermal endothelial cells and keratinocytes [83–85], whereas CCL22 is secreted by macrophages, interstitial dendritic cells, and epidermal Langerhans cells [86, 87]. Hence, CCL17 expressed by the dermal endothelial cells and infiltrating dermal cells of atopic lesional skin may act in the first steps of T cell recruitment by inducing integrin-dependent adhesion and transendothelial migration of T cells while CCL22 supports the formation of T cell-dendritic cell clusters at sites of atopic skin inflammation.

A recent study in mice by Reiss et al. suggests that ligands of CCR4 and CCR10 cooperate in the recruitment of memory T cells to sites of skin inflammation [88]. According to this model, CCL17 displayed by cutaneous venules, in combination with other CCR4 ligands, trigger the integrin-dependent arrest and extravasation of lymphocytes rolling on cutaneous venules. Subsequently, CCL27, highly and selectively expressed by keratinocytes, may support diapedesis and epidermotropism of skin homing T cells [88, 89].

A systematical analysis of the expression of all known chemokines in chronic inflammatory skin diseases identified CCL18/*DC-CK1*/*PARC* to be specifically associated with an atopic dermatitis phenotype but absent in other chronic inflammatory or autoimmune skin diseases such as psoriasis or cutaneous lupus erythematosus [90]. Among all known chemokines, CCL18 represented the most highly expressed ligand in atopic dermatitis and the absolute amount of CCL18 mRNA in lesional atopic skin was more than 100-fold higher than those seen for CCL17 [90]. In good accordance with this finding, a DNA microarray screen also identified CCL18 as one of the genes showing the strongest association with atopic dermatitis compared to psoriatic or normal skin specimen [91]. Interestingly, trigger factors of atopic skin inflammation, such as allergen exposure and staphylococcal superantigens markedly induced this chemokine *in vitro* and *in vivo* suggesting important CCL18-driven processes during the initiation and amplification of atopic skin inflammation [90]. Although its receptor is yet unidentified, CCL18 binds to CLA^+ skin-homing memory T cells and induces the migration of memory T cells into the human skin, *in vivo* [92]. CCL18 is produced by dermal dendritic cells in close proximity to infiltrating T cells implicating a role in the formation of T cell-dendritic cell contacts within atopic skin [90].

Psoriasis vulgaris

Psoriasis vulgaris represents a common chronically relapsing inflammatory skin disease affecting approximately 1–2% of the general population [93, 94]. Psoriatic patients suffer from erythemato-squamous plaques predominantly manifesting at the extensor parts of joints, above the Os sacrum and the capillitium [93, 94]. In severe cases, skin lesions can involve the entire integument and be accompanied by a destructive psoriatic arthritis. Histopathologically, psoriasis is characterised by a marked inflammatory infiltrate, hyperproliferation of keratinocytes, elongation of rete ridges and hyperconvuluted vascular corpores in the papillary dermis [93, 94]. The infiltrate is composed of skin-infiltrating CLA^+ memory T cells predominantly showing a Th1 phenotype, neutrophils, lining macrophages and increased numbers of dendritic cells. There is evidence that T cells play a crucial role in the immunopathogenesis of this disease [1, 75, 77]. An early cellular event in the development of psoriatic lesions is the infiltration of target sites by activated T cells, which in turn produce inflammatory mediators, such as IFN-γ, induce epidermal hyperplasia and may act with keratinocytes and dermal macrophages to sustain a cycle of inflammation which finally leads to the psoriatic phenotype [95].

To date, there are no studies showing the efficacy of therapeutic targeting of chemokine ligand-receptor interactions in mouse models for psoriasis. However, there is increasing knowledge of chemokines and chemokine receptors associated with a psoriasis phenotype (Tab. 1).

One such example represents CXCL8/IL-8. CXCL8 was initially identified in and extracted from psoriatic scales and probably represents one of the most intensively characterised chemokines known so far. Although CXCL8 was already identified 17 years ago, the investigation of its functional role *in vivo* had been limited since there exists no orthologue in the mouse.

Recently, Rottman et al. suggested a potential role for CXCR3 and CCR4 ligands in the pathogenesis of psoriasis [96]. CXCR3 and CCR4 were expressed on $CD3^+$ dermal lymphocytes and chemokine receptor expression was accompanied by the up-regulation of their respective ligands, CXCL9 and CXCL10 as well as CCL17 and CCL22 in lesional psoriatic skin. Furthermore, TNF-α and IFN-γ were identified to regulate those psoriasis-associated genes in keratinocytes and dermal endothelial cells. In contrast to skin-infiltrating dermal lymhocytes, epidermal lymphocyte subsets were characterised by the co-expression of CLA, $\alpha_E\beta_7$ and CXCR3 while CCR4 was absent. The authors suggest a model with CCR4 and CXCR3 ligands mediating tethering and transendothelial migration of CLA^+ T cells and subsequent involvement of CXCR3 ligands in directing lymophocytes into the epidermis. During this migration process the adhesion molecule $\alpha_E\beta_7$ may be up-regulated through dermal fibroblast-derived TGF-β stimulation and support the anchoring of epidermis infiltrating lymphocytes by its heterotypic interaction with E-cadherin on keratinocytes [96].

Although the significance of inflammatory chemokines to lymphocyte recruitment *in vivo* remains unclear, another inflammatory chemokine, CCL20, shows an interesting association with a psoriatic phenotype [66]. CCL20 is known to attract both T and dendritic cells [66, 97, 98]. Among dendritic cells, CCL20 is a highly potent chemokine for the chemoattraction of epithelial Langerhans-type dendritic cells [97, 99]. Furthermore, CCL20 has been shown to preferentially attract the memory subset of T cells [98]. This CC chemokine and its receptor CCR6 are significantly up-regulated in psoriatic skin [66]. Within psoriatic lesions, CCL20-expressing keratinocytes co-localise with skin-infiltrating T lymphocytes. Furthermore, CCR6 is expressed at high levels on the skin-homing CLA^+ subset of memory T cells [66]. Psoriatic skin-homing CLA^+ T cells show increased chemotactic responses towards CCL20 gradients when compared to those of normal donors [66]. TNF-α and IL-1, both pro-inflammatory cytokines known to be up-regulated in psoriasis, as well as CD40L are potent inducers of bioactive CCL20 protein in keratinocytes, dermal microvascular endothelial cells, dermal fibroblast and dendritic cells *in vitro* [66]. Furthermore, T helper cell-derived mediators (e.g., IFN-γ, IL-17, CD40L) regulate CCL20 production in cellular constituents of the skin. IL-17 is known to be up-regulated in lesional psoriatic skin, suggesting that it may play a role in the amplification and/or development of cutaneous inflammation [66]. Along with its expression in intestinal epithelial cells, cutaneous CCL20 expression supports the hypothesis that this inflammatory chemokine plays an important role in the interface between the host and the environment.

Generation of intestinal and skin tropic effector T cell subsets

Recent studies in mice, examining cell adhesion molecule and chemokine receptor expression on adoptively transferred TCR transgenic T cells, have demonstrated a critical role for the local draining lymph nodes in the generation of 'tissue tropic' effector T cell subsets. Thus, T cells activated in MLN were induced to express the 'gut tropic' markers $\alpha_4\beta_7$ and CCR9, while T cells activated in skin draining LN were induced to express E-selectin ligands [21, 100]. DCs isolated from the MLN and PP were necessary and sufficient for the induction of $\alpha_4\beta_7$ and CCR9 on responding T cells *in vitro*, and both $CD8^+$ and $CD8^-$ MLN DC, and $CD8^+$ depleted PP DCs could generate gut tropic effector T cells [28, 101]. In contrast priming with Langerhan cells from the skin lead to a dramatic induction of CCR4 and E-selectin ligands on responding cells (Fig. 3, [28, 101–103]). Together these results demonstrate a critical role for environmentally imprinted DCs, in the generation of tissue tropic effector T cell subsets. Whether DCs in skin draining LN induce other chemokine receptors associated with skin tropic T cells such as CCR8 and CCR10 has yet to be determined. Reactivation of tissue tropic memory T cell subsets can modify their tissue tropism according to the origin of the last activating DC. In this way gut tropic memory T cells can be reprogrammed to express markers of skin tropism and *vice versa* [104]. The underlying mechanism by which environmentally imprinted DCs generate tissue tropic effector T cell subsets is poorly understood. Recently however, the vitamin A metabolite, retinoic acid was found to induce $\alpha_4\beta_7$ and CCR9 on *in vitro* activated T cells and to suppress expression of skin homing markers [105]. Importantly, MLN and PP DCs, but not spleen DCs, could produce retinoic acid, and their ability to generate gut tropic T cells was reduced with an inhibitor to enzymes involved in retinoic acid synthesis as well as an antagonist to retinoic acid receptors [105].

Conclusions

Peripheral epithelial tissues are a rich source of chemokines under both homeostatic and inflammatory conditions. Tissue tropism (e.g., skin *versus* gut as discussed here) on effector T cell subsets is imprinted by environmentally modulated DCs within local LNs and involves the selective induction of specific chemokine receptors. These tissue-selective chemokine systems, together with appropriate adhesion molecules, are essential regulators of effector lymphocyte trafficking to peripheral sites. Dysregulated lymphocyte accumulation and activation appears to be an important driving factor for chronic inflammation at these sites; therefore, targeting the chemokine pathway(s) to block lymphocyte infiltration may provide a means for alleviating disease symptoms. Nevertheless, many questions must be answered before chemokine/chemokine receptors can be chosen as novel targets for the treatment of intestinal and skin inflammation. A clearer picture of the role of individual chemokines in animal models of inflammation will be a critical step before singling out certain chemokines or their chemokine receptors as novel drug targets. Such studies should help to determine whether neutralisation of a single chemokine sys-

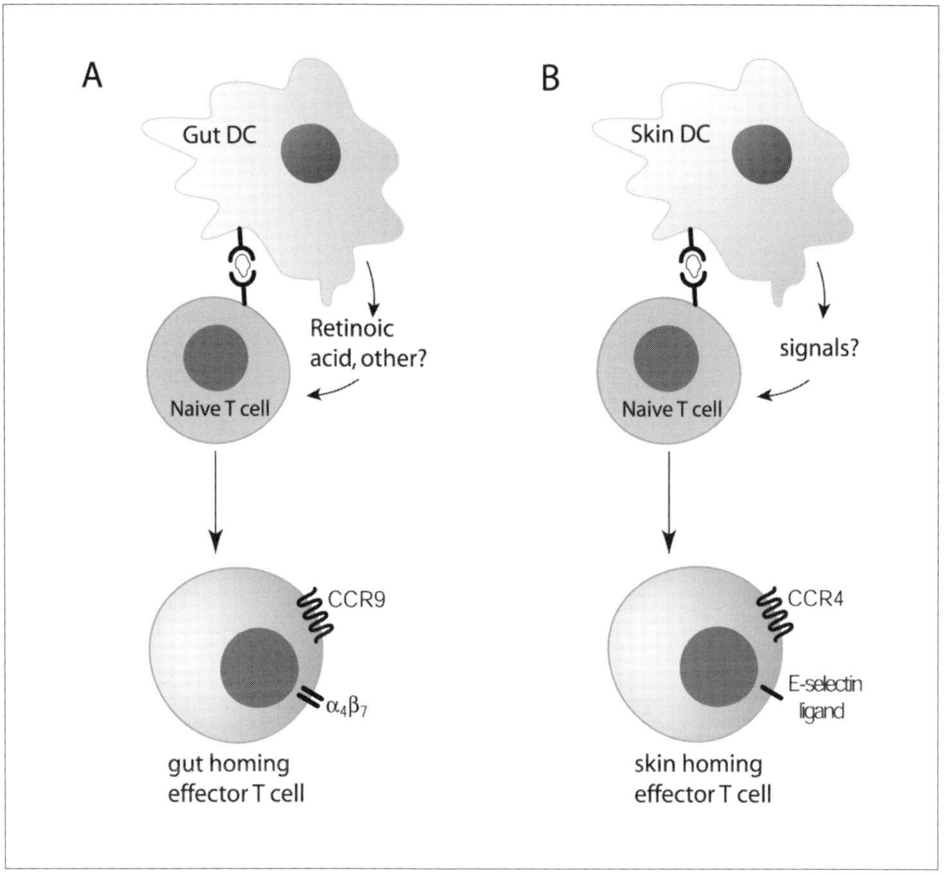

Figure 3
Dendritic cells play a critical role in the generation of tissue tropic effector T cell subsets (A) MLN or PP dendritic cells (DC) generate $CCR9^+\alpha_4\beta_7^+$ gut homing effector T cells [28, 101]. This ability appears, at least in part, to be due to their ability to generate retinoic acid [105]. (B) T cells activated by skin dendritic cells (Langerhans cells) are induced to express E-selectin ligand and CCR4 [106], molecules implicated in T cell homing to the inflamed skin.

tem is sufficient to treat inflammatory skin/gut diseases or whether it will be necessary to target multiple chemokine receptors in combination. Also, we need to learn more about the kinetics of disease progression and the time point for optimal interference. For example, it is important to know if a given chemokine system is primarily involved in the acute phase of disease as opposed to the chronic stage of disease or disease recurrence. Finally, identifying the factors within peripheral tissues and draining LNs that determine tissue tropism on newly generated effector T cells are likely to provide novel strategies for interference with lymphocyte trafficking to peripheral epithelial tissues.

Acknowledgements
We would like to thank Dr J Marsal for his help with Figure 1. This work was supported by grants from the Swedish Science Medical Research Council, the Crafoordska, Österlund, Åke Wiberg, Richard and Ruth Julins, Nanna Svartz and Kocks foundations, the Royal Physiographic Society, the Swedish foundation for Strategic Research "Microbes and Man" and INGVAR programs, the German Research Foundation (SFB503/C9) and the European Union (QLRK4-CT-2001-00366).

References

1 Lefrancois L, Parker CM, Olson S, Muller W, Wagner N, Schon MP, Puddington L (1999) The role of beta7 integrins in CD8 T cell trafficking during an antiviral immune response. *J Exp Med* 189: 1631–1638
2 Berlin C, Bargatze RF, Campbell JJ, von Andrian UH, Szabo MC, Hasslen SR, Nelson RD, Berg EL, Erlandsen SL, Butcher EC (1995) Alpha 4 integrins mediate lymphocyte attachment and rolling under physiologic flow. *Cell* 80: 413–422
3 Hamann A, Andrew DP, Jablonski-Westrich D, Holzmann B, Butcher EC (1994) Role of alpha 4-integrins in lymphocyte homing to mucosal tissues *in vivo*. *J Immunol* 152: 3282–3293
4 Rott LS, Rose JR, Bass D, Williams MB, Greenberg HB, Butcher EC (1997) Expression of mucosal homing receptor alpha4beta7 by circulating $CD4^+$ cells with memory for intestinal rotavirus. *J Clin Invest* 100: 1204–1208
5 Hayday A, Theodoridis E, Ramsburg E, Shires J (2001) Intraepithelial lymphocytes: exploring the Third Way in immunology. *Nat Immunol* 2: 997–1003
6 Cheroutre H (2004) Starting at the beginning: new perspectives on the biology of mucosal T cells. *Annu Rev Immunol* 22: 217–246
7 Agace W, Roberts A, Wu L, Greineder C, Ebert E, Parker C (2000) Human intestinal lamina propria and intraepithelial lymphocytes express receptors specific for chemokines induced by inflammation. *Eur J Immunol* 30: 819–826
8 Zabel BA, Agace WW, Campbell JJ, Heath HM, Parent D, Roberts AI, Ebert EC, Kassam N, Qin S, Zovko M et al (1999) Human G Protein-coupled receptor GPR-9-6/CC chemokine receptor 9 is selectively expressed on intestinal homing T lymphocytes, mucosal lymphocytes, and thymocytes and is required for thymus-expressed chemokine-mediated chemotaxis. *J Exp Med* 190: 1241–1256
9 Agace W, Amara A, Roberts A, Pablos J, Thelen S, Uguccioni M, Marsal J, Arenzana-Seisdedos F, Delaunay T, Ebert E et al (2000) Constitutive expression of stromal derived

factor-1 by mucosal epithelia and its role in HIV transmission and propagation. *Curr Biol* 10: 325–328
10 Kunkel EJ, Kim CH, Lazarus NH, Vierra MA, Soler D, Bowman EP, Butcher EC (2003) CCR10 expression is a common feature of circulating and mucosal epithelial tissue IgA Ab-secreting cells. *J Clin Invest* 111: 1001–1010
11 Schaerli P, Ebert L, Willimann K, Blaser A, Roos RS, Loetscher P, Moser B (2004) A skin-selective homing mechanism for human immune surveillance T cells. *J Exp Med* 199: 1265–1275
12 Lapenta C, Boirivant M, Marini M, Santini SM, Logozzi M, Viora M, Belardelli F, Fais S (1999) Human intestinal lamina propria lymphocytes are naturally permissive to HIV-1 infection. *Eur J Immunol* 29: 1202–1208
13 Lugering A, Kucharzik T, Soler D, Picarella D, Hudson JT 3rd, Williams IR (2003) Lymphoid precursors in intestinal cryptopatches express CCR6 and undergo dysregulated development in the absence of CCR6. *J Immunol* 171: 2208–2215
14 Ericsson A, Svensson M, Arya A, Agace WW (2004) CCL25/CCR9 promotes the induction and function of CD103 on intestinal intraepithelial lymphocytes. *Eur J Immunol* 34: 2720–2729
15 Matloubian M, David A, Engel S, Ryan JE, Cyster JG (2000) A transmembrane CXC chemokine is a ligand for HIV-coreceptor Bonzo. *Nat Immunol* 1: 298–304
16 Unutmaz D, Xiang W, Sunshine MJ, Campbell J, Butcher E, Littman DR (2000) The primate lentiviral receptor Bonzo/STRL33 is coordinately regulated with CCR5 and its expression pattern is conserved between human and mouse. *J Immunol* 165: 3284–3292
17 Connor SJ, Paraskevopoulos N, Newman R, Cuan N, Hampartzoumian T, Lloyd AR, Grimm MC (2004) CCR2 expressing $CD4^+$ T lymphocytes are preferentially recruited to the ileum in Crohn's disease. *Gut* 53: 1287–1294
18 Kunkel EJ, Boisvert J, Murphy K, Vierra MA, Genovese MC, Wardlaw AJ, Greenberg HB, Hodge MR, Wu L, Butcher EC et al (2002) Expression of the chemokine receptors CCR4, CCR5, and CXCR3 by human tissue-infiltrating lymphocytes. *Am J Pathol* 160: 347–355
19 Kunkel EJ, Campbell JJ, Haraldsen G, Pan J, Boisvert J, Roberts AI, Ebert EC, Vierra M A, Goodman SB, Genovese MC et al (2000) Lymphocyte CC chemokine receptor 9 and epithelial thymus-expressed chemokine (TECK) expression distinguish the small intestinal immune compartment: Epithelial expression of tissue-specific chemokines as an organizing principle in regional immunity. *J Exp Med* 192: 761–768
20 Papadakis KA, Prehn J, Nelson V, Cheng L, Binder SW, Ponath PD, Andrew DP, Targan SR (2000) The role of thymus-expressed chemokine and its receptor CCR9 on lymphocytes in the regional specialization of the mucosal immune system. *J Immunol* 165: 5069–5076
21 Svensson M, Marsal J, Ericsson A, Carramolino L, Broden T, Marquez G, Agace WW (2002) CCL25 mediates the localization of recently activated CD8alphabeta($^+$) lymphocytes to the small-intestinal mucosa. *J Clin Invest* 110: 1113–1121
22 Wurbel MA, Philippe JM, Nguyen C, Victorero G, Freeman T, Wooding P, Miazek A,

Mattei MG, Malissen M, Jordan BR et al (2000) The chemokine TECK is expressed by thymic and intestinal epithelial cells and attracts double- and single-positive thymocytes expressing the TECK receptor CCR9 [In Process Citation]. *Eur J Immunol* 30: 262–271

23 Wurbel MA, Malissen M, Guy-Grand D, Meffre E, Nussenzweig MC, Richelme M, Carrier A, Malissen B (2001) Mice lacking the CCR9 CC-chemokine receptor show a mild impairment of early T- and B-cell development and a reduction in T-cell receptor gammadelta(+) gut intraepithelial lymphocytes. *Blood* 98: 2626–2632

24 Uehara S, Grinberg A, Farber JM, Love PE (2002) A role for CCR9 in T lymphocyte development and migration. *J Immunol* 168: 2811–2819

25 Pabst O, Ohl L, Wendland M, Wurbel MA, Kremmer E, Malissen B, Forster R (2004) Chemokine receptor CCR9 contributes to the localization of plasma cells to the small intestine. *J Exp Med* 199: 411–416. Epub 26 January 2004

26 Marsal J, Svensson M, Ericsson A, Iranpour AH, Carramolino L, Marquez G, Agace WW (2002) Involvement of CCL25 (TECK) in the generation of the murine small-intestinal CD8alpha alpha$^+$CD3$^+$ intraepithelial lymphocyte compartment. *Eur J Immunol* 32: 3488–3497

27 Onai N, Kitabatake M, Zhang YY, Ishikawa H, Ishikawa S, Matsushima K (2002) Pivotal role of CCL25 (TECK)-CCR9 in the formation of gut cryptopatches and consequent appearance of intestinal intraepithelial T lymphocytes. *Int Immunol* 14: 687–694

28 Johansson-Lindbom B, Svensson M, Wurbel M-A, Malissen B, Márquez G, Agace W (2003) Selective generation of gut-tropic T cells in gut associated lymphoid tissue (GALT); requirement for GALT dendritic cells and adjuvant. *J Exp Med* 198: 963–969

29 Hieshima K, Kawasaki Y, Hanamoto H, Nakayama T, Nagakubo D, Kanamaru A, Yoshie O (2004) CC chemokine ligands 25 and 28 play essential roles in intestinal extravasation of IgA antibody-secreting cells. *J Immunol* 173: 3668–3675

30 Hosoe N, Miura S, Watanabe C, Tsuzuki Y, Hokari R, Oyama T, Fujiyama Y, Nagata H, Ishii H (2004) Demonstration of functional role of TECK/CCL25 in T lymphocyte-endothelium interaction in inflamed and uninflamed intestinal mucosa. *Am J Physiol Gastrointest Liver Physiol* 286: G458–G466

31 Dwinell MB, Lugering N, Eckmann L, Kagnoff MF (2001) Regulated production of interferon-inducible T-cell chemoattractants by human intestinal epithelial cells. *Gastroenterology* 120: 49–59

32 Shibahara T, Wilcox JN, Couse T, Madara JL (2001) Characterization of epithelial chemoattractants for human intestinal intraepithelial lymphocytes. *Gastroenterology* 120: 60–70

33 Grimm M, Doe W (1996) Chemokines in inflammatory bowel disease mucosa: expression of RANTES, Macrophage Inflammatory Protein (MIP)-1α, MIP-1β, and γ-interferon-inducible protein-10 by macrophages, lymphocytes, endothelial cells and granulomas. *Inflam Bow Dis* 2: 88–96

34 Uguccioni M, Gionchetti P, Robbiani DF, Rizzello F, Peruzzo S, Campieri M, Baggiolini M (1999) Increased expression of IP-10, IL-8, MCP-1, and MCP-3 in ulcerative colitis. *Am J Pathol* 155: 331–336

35 Mazzucchelli L, Hauser C, Zgraggen K, Wagner HE, Hess MW, Laissue JA, Mueller C (1996) Differential *in situ* expression of the genes encoding the chemokines MCP-1 and RANTES in human inflammatory bowel disease. *J Pathol* 178: 201–206

36 Tanaka Y, Imai T, Baba M, Ishikawa I, Uehira M, Nomiyama H, Yoshie O (1999) Selective expression of liver and activation-regulated chemokine (LARC) in intestinal epithelium in mice and humans. *Eur J Immunol* 29: 633–642

37 Cook DN, Prosser DM, Forster R, Zhang J, Kuklin NA, Abbondanzo SJ, Niu XD, Chen SC, Manfra DJ, Wiekowski MT et al (2000) CCR6 mediates dendritic cell localization, lymphocyte homeostasis, and immune responses in mucosal tissue. *Immunity* 12: 495–503

38 Varona R, Villares R, Carramolino L, Goya I, Zaballos A, Gutierrez J, Torres M, Martinez AC, Marquez G (2001) CCR6-deficient mice have impaired leukocyte homeostasis and altered contact hypersensitivity and delayed-type hypersensitivity responses. *J Clin Invest* 107: R37–R45

39 Zhao X, Sato A, Dela Cruz CS, Linehan M, Luegering A, Kucharzik T, Shirakawa AK, Marquez G, Farber JM, Williams I et al (2003) CCL9 is secreted by the follicle-associated epithelium and recruits dome region Peyer's patch $CD11b^+$ dendritic cells. *J Immunol* 171: 2797–2803

40 Banks C, Bateman A, Payne R, Johnson P, Sheron N (2003) Chemokine expression in IBD. Mucosal chemokine expression is unselectively increased in both ulcerative colitis and Crohn's disease. *J Pathol* 199: 28–35

41 Wedemeyer J, Lorentz A, Goke M, Meier PN, Flemming P, Dahinden CA, Manns MP, Bischoff SC (1999) Enhanced production of monocyte chemotactic protein 3 in inflammatory bowel disease mucosa [In Process Citation]. *Gut* 44: 629–635

42 Grimm MC, Elsbury SK, Pavli P, Doe WF (1996) Enhanced expression and production of monocyte chemoattractant protein-1 in inflammatory bowel disease mucosa. *J Leukoc Biol* 59: 804–812

43 Reinecker HC, Loh EY, Ringler DJ, Mehta A, Rombeau JL, MacDermott RP (1995) Monocyte-chemoattractant protein 1 gene expression in intestinal epithelial cells and inflammatory bowel disease mucosa. *Gastroenterology* 108: 40–50

44 DePaolo RW, Rollins BJ, Kuziel W, Karpus WJ (2003) CC chemokine ligand 2 and its receptor regulate mucosal production of IL-12 and TGF-beta in high dose oral tolerance. *J Immunol* 171: 3560–3567

45 Heidemann J, Ogawa H, Rafiee P, Lugering N, Maaser C, Domschke W, Binion DG, Dwinell MB (2004) Mucosal angiogenesis regulation by CXCR4 and its ligand CXCL12 expressed by human intestinal microvascular endothelial cells. *Am J Physiol Gastrointest Liver Physiol* 286: G1059–G1068. Epub 05 February 2004

46 Wright N, Hidalgo A, Rodriguez-Frade JM, Soriano SF, Mellado M, Parmo-Cabanas M, Briskin MJ, Teixido J (2002) The chemokine stromal cell-derived factor-1 alpha modulates alpha 4 beta 7 integrin-mediated lymphocyte adhesion to mucosal addressin cell adhesion molecule-1 and fibronectin. *J Immunol* 168: 5268–5277

47 Bowman EP, Kuklin NA, Youngman KR, Lazarus NH, Kunkel EJ, Pan J, Greenberg HB,

Butcher EC (2002) The intestinal chemokine thymus-expressed chemokine (CCL25) attracts IgA antibody-secreting cells. *J Exp Med* 195: 269–275

48 Wang W, Soto H, Oldham ER, Buchanan ME, Homey B, Catron D, Jenkins N, Copeland NG, Gilbert DJ, Nguyen N et al (2000) Identification of a novel chemokine (CCL28), which binds CCR10 (GPR2). *J Biol Chem* 275: 22313–22323

49 Pan J, Kunkel EJ, Gosslar U, Lazarus N, Langdon P, Broadwell K, Vierra MA, Genovese MC, Butcher EC, Soler D (2000) A novel chemokine ligand for CCR10 and CCR3 expressed by epithelial cells in mucosal tissues. *J Immunol* 165: 2943–2949

50 Oki M, Ohtani H, Kinouchi Y, Sato E, Nakamura S, Matsumoto T, Nagura H, Yoshie O, Shimosegawa T (2005) Accumulation of $CCR5^+$ T cells around $RANTES^+$ granulomas in Crohn's disease: a pivotal site of Th1-shifted immune response? *Lab Invest* 85: 137–145

51 Kwon JH, Keates S, Bassani L, Mayer LF, Keates AC (2002) Colonic epithelial cells are a major site of macrophage inflammatory protein 3alpha (MIP-3alpha) production in normal colon and inflammatory bowel disease. *Gut* 51: 818–826

52 Kaser A, Ludwiczek O, Holzmann S, Moschen AR, Weiss G, Enrich B, Graziadei I, Dunzendorfer S, Wiedermann CJ, Murzl E et al (2004) Increased expression of CCL20 in human inflammatory bowel disease. *J Clin Immunol* 24: 74–85

53 Papadakis KA, Prehn J, Moreno ST, Cheng L, Kouroumalis EA, Deem R, Breaverman T, Ponath PD, Andrew DP, Green PH et al (2001) CCR9-positive lymphocytes and thymus-expressed chemokine distinguish small bowel from colonic Crohn's disease. *Gastroenterology* 121: 246–254

54 Ajuebor MN, Hogaboam CM, Kunkel SL, Proudfoot AE, Wallace JL (2001) The chemokine RANTES is a crucial mediator of the progression from acute to chronic colitis in the rat. *J Immunol* 166: 552–558

55 Varona R, Cadenas V, Flores J, Martinez AC, Marquez G (2003) CCR6 has a nonredundant role in the development of inflammatory bowel disease. *Eur J Immunol* 33: 2937–2946

56 Sasaki S, Yoneyama H, Suzuki K, Suriki H, Aiba T, Watanabe S, Kawauchi Y, Kawachi H, Shimizu F, Matsushima K et al (2002) Blockade of CXCL10 protects mice from acute colitis and enhances crypt cell survival. *Eur J Immunol* 32: 3197–3205

57 Singh UP, Singh S, Taub DD, Lillard JW Jr (2003) Inhibition of IFN-gamma-inducible protein-10 abrogates colitis in $IL\text{-}10^{-/-}$ mice. *J Immunol* 171: 1401–1406

58 Andres PG, Beck PL, Mizoguchi E, Mizoguchi A, Bhan AK, Dawson T, Kuziel WA, Maeda N, MacDermott RP, Podolsky DK et al (2000) Mice with a selective deletion of the CC chemokine receptors 5 or 2 are protected from dextran sodium sulfate-mediated colitis: lack of CC chemokine receptor 5 expression results in a $NK1.1^+$ lymphocyte-associated Th2-type immune response in the intestine. *J Immunol* 164: 6303–6312

59 Luangsay S, Kasper LH, Rachinel N, Minns LA, Mennechet FJ, Vandewalle A Buzoni-Gatel D (2003) CCR5 mediates specific migration of *Toxoplasma gondii*-primed CD8 lymphocytes to inflammatory intestinal epithelial cells. *Gastroenterology* 125: 491–500

60 Picker LJ (1993) Regulation of tissue-selective T-lymphocyte homing receptors during

the virgin to memory/effector cell transition in human secondary lymphoid tissues. *Am Rev Respir Dis* 148: S47–S54

61 Picker LJ, Martin RJ, Trumble A, Newman LS, Collins PA, Bergstresser PR, Leung DY (1994) Differential expression of lymphocyte homing receptors by human memory/effector T cells in pulmonary *versus* cutaneous immune effector sites. *Eur J Immunol* 24: 1269–1277

62 Picker LJ, Terstappen LW, Rott LS, Streeter PR, Stein H, Butcher EC (1990) Differential expression of homing-associated adhesion molecules by T cell subsets in man. *J Immunol* 145: 3247–3255

63 Santamaria Babi LF, Moser R, Perez Soler MT, Picker LJ, Blaser K, Hauser C (1995) Migration of skin-homing T cells across cytokine-activated human endothelial cell layers involves interaction of the cutaneous lymphocyte-associated antigen (CLA), the very late antigen-4 (VLA-4), and the lymphocyte function-associated antigen-1 (LFA-1). *J Immunol* 154: 1543–1550

64 Homey B, Alenius H, Muller A, Soto H, Bowman EP, Yuan W, McEvoy L, Lauerma AI, Assmann T, Bunemann E et al (2002) CCL27-CCR10 interactions regulate T cell-mediated skin inflammation. *Nat Med* 8: 157–165

65 Caux C, Ait-Yahia S, Chemin K, de Bouteiller O, Dieu-Nosjean MC, Homey B, Massacrier C, Vanbervliet B, Zlotnik A, Vicari A (2000) Dendritic cell biology and regulation of dendritic cell trafficking by chemokines. *Springer Semin Immunopathol* 22: 345–369

66 Homey B, Dieu-Nosjean MC, Wiesenborn A, Massacrier C, Pin JJ, Oldham E, Catron D, Buchanan ME, Muller A, deWaal Malefyt R et al (2000) Up-regulation of macrophage inflammatory protein-3 alpha/CCL20 and CC chemokine receptor 6 in psoriasis. *J Immunol* 164: 6621–6632

67 Hudak S, Hagen M, Liu Y, Catron D, Oldham E, McEvoy LM, Bowman EP (2002) Immune surveillance and effector functions of CCR10($^+$) skin homing T cells. *J Immunol* 169: 1189–1196

68 Soler D, Humphreys TL, Spinola SM, Campbell JJ (2003) CCR4 *versus* CCR10 in human cutaneous TH lymphocyte trafficking. Blood 101: 1677–1682. Epub 24 October 2002

69 Charbonnier AS, Kohrgruber N, Kriehuber E, Stingl G, Rot A, Maurer D (1999) Macrophage inflammatory protein 3alpha is involved in the constitutive trafficking of epidermal Langerhans cells. *J Exp Med* 190: 1755–1768

70 Morales J, Homey B, Vicari AP, Hudak S, Oldham E, Hedrick J, Orozco R, Copeland NG, Jenkins NA, McEvoy LM et al (1999) CTACK, a skin-associated chemokine that preferentially attracts skin-homing memory T cells. *Proc Natl Acad Sci USA* 96: 14470–14475

71 Homey B, Wang W, Soto H, Buchanan ME, Wiesenborn A, Catron D, Muller A, McClanahan TK, Dieu-Nosjean MC, Orozco R et al (2000) Cutting edge: the orphan chemokine receptor G protein-coupled receptor-2 (GPR-2, CCR10) binds the skin-associated chemokine CCL27 (CTACK/ALP/ILC). *J Immunol* 164: 3465–3470

72 Baekkevold ES, Yamanaka T, Palframan RT, Carlsen HS, Reinholt FP, von Andrian UH, Brandtzaeg P, Haraldsen G (2001) The CCR7 ligand elc (CCL19) is transcytosed in high endothelial venules and mediates T cell recruitment. *J Exp Med* 193: 1105–1112

73 Middleton J, Neil S, Wintle J, Clark-Lewis I, Moore H, Lam C, Auer M, Hub E, Rot A (1997) Transcytosis and surface presentation of IL-8 by venular endothelial cells. *Cell* 91: 385–395

74 Leung DY, Bieber T (2003) Atopic dermatitis. *Lancet* 361: 151–160

75 Nickoloff BJ, Wrone-Smith T (1999) Injection of pre-psoriatic skin with $CD4^+$ T cells induces psoriasis. *Am J Pathol* 155: 145–158

76 Schon MP (1999) Animal models of psoriasis – what can we learn from them? *J Invest Dermatol* 112: 405–410

77 Schon MP, Detmar M, Parker CM (1997) Murine psoriasis-like disorder induced by naive $CD4^+$ T cells. *Nat Med* 3: 183–188

78 Grewe M, Bruijnzeel-Koomen CA, Schopf E, Thepen T, Langeveld-Wildschut AG, Ruzicka T, Krutmann J (1998) A role for Th1 and Th2 cells in the immunopathogenesis of atopic dermatitis. *Immunol Today* 19: 359–361

79 Biedermann T, Schwarzler C, Lametschwandtner G, Thoma G, Carballido-Perrig N, Kund J, de Vries JE, Rot A, Carballido JM (2002) Targeting CLA/E-selectin interactions prevents CCR4-mediated recruitment of human Th2 memory cells to human skin *in vivo*. *Eur J Immunol* 32: 3171–3180

80 Campbell JJ, Haraldsen G, Pan J, Rottman J, Qin S, Ponath P, Andrew DP, Warnke R, Ruffing N, Kassam N et al (1999) The chemokine receptor CCR4 in vascular recognition by cutaneous but not intestinal memory T cells. *Nature* 400: 776–780

81 Imai T, Baba M, Nishimura M, Kakizaki M, Takagi S, Yoshie O (1997) The T cell-directed CC chemokine TARC is a highly specific biological ligand for CC chemokine receptor 4. *J Biol Chem* 272: 15036–15042

82 Zlotnik A, Yoshie O (2000) Chemokines: a new classification system and their role in immunity. *Immunity* 12: 121–127

83 Vestergaard C, Bang K, Gesser B, Yoneyama H, Matsushima K, Larsen CG, Murai M, Nakamura K, Tamaki K, Terashima Y et al (2000) A Th2 chemokine, TARC, produced by keratinocytes may recruit CLA^+CCR4^+ lymphocytes into lesional atopic dermatitis skin; Overproduction of Th2-specific chemokines in NC/Nga mice exhibiting atopic dermatitis-like lesions. *J Invest Dermatol* 115: 640–646

84 Vestergaard C, Deleuran M, Gesser B, Gronhoj Larsen C, Bang K, Yoneyama H, Matsushima K, Larsen CG, Murai M, Nakamura K et al (2003) Expression of the T-helper 2-specific chemokine receptor CCR4 on CCR10-positive lymphocytes in atopic dermatitis skin but not in psoriasis skin. *Br J Dermatol* 149: 457–463

85 Zheng X, Nakamura K, Furukawa H, Nishibu A, Takahashi M, Tojo M, Kaneko F, Kakinuma T, Tamaki K (2003) Demonstration of TARC and CCR4 mRNA expression and distribution using *in situ* RT-PCR in the lesional skin of atopic dermatitis. *J Dermatol* 30: 26–32

86 Katou F, Ohtani H, Nakayama T, Ono K, Matsushima K, Saaristo A, Nagura H, Yoshie

O, Motegi K (2001) Macrophage-derived chemokine (MDC/CCL22) and CCR4 are involved in the formation of T lymphocyte-dendritic cell clusters in human inflamed skin and secondary lymphoid tissue. *Am J Pathol* 158: 1263–1270
87 Vulcano M, Albanesi C, Stoppacciaro A, Bagnati R, D'Amico G, Struyf S, Transidico P, Bonecchi R, Del Prete A, Allavena P et al (2001) Dendritic cells as a major source of macrophage-derived chemokine/CCL22 *in vitro* and *in vivo*. *Eur J Immunol* 31: 812–822
88 Reiss Y, Proudfoot AE, Power CA, Campbell JJ, Butcher EC (2001) CC chemokine receptor (CCR)4 and the CCR10 ligand cutaneous T cell-attracting chemokine (CTACK) in lymphocyte trafficking to inflamed skin. *J Exp Med* 194: 1541–1547
89 Campbell JJ, Butcher EC (2000) Chemokines in tissue-specific and microenvironment-specific lymphocyte homing. *Curr Opin Immunol* 12: 336–341
90 Pivarcsi A, Gombert M, Dieu-Nosjean MC, Lauerma A, Kubitza R, Meller S, Rieker J, Muller A, Da Cunha L, Haahtela A et al (2004) CC chemokine ligand 18, an atopic dermatitis-associated and dendritic cell-derived chemokine, is regulated by staphylococcal products and allergen exposure. *J Immunol* 173: 5810–5817
91 Nomura I, Gao B, Boguniewicz M, Darst MA, Travers JB, Leung DY (2003) Distinct patterns of gene expression in the skin lesions of atopic dermatitis and psoriasis: a gene microarray analysis. *J Allergy Clin Immunol* 112: 1195–1202
92 Gunther C, Bello-Fernandez C, Kopp T, Kund J, Carballido-Perrig N, Hinteregger S, Fassl S, Schwarzler C, Lametschwandtner G, Stingl G et al (2005) CCL18 is expressed in atopic dermatitis and mediates skin homing of human memory T cells. *J Immunol* 174: 1723–1728
93 Elder JT, Nair RP, Voorhees JJ (1994) Epidemiology and the genetics of psoriasis. *J Invest Dermatol* 102: 24S–27S
94 Ortonne JP (1996) Aetiology and pathogenesis of psoriasis. *Br J Dermatol* 135 (Suppl 49): 1–5
95 Bata-Csorgo Z, Hammerberg C, Voorhees JJ, Cooper KD (1995) Intralesional T-lymphocyte activation as a mediator of psoriatic epidermal hyperplasia. *J Invest Dermatol* 105: 89S–94S
96 Rottman JB, Smith TL, Ganley KG, Kikuchi T, Krueger JG (2001) Potential role of the chemokine receptors CXCR3, CCR4, and the integrin alphaEbeta7 in the pathogenesis of psoriasis vulgaris. *Lab Invest* 81: 335–347
97 Dieu-Nosjean MC, Massacrier C, Homey B, Vanbervliet B, Pin JJ, Vicari A, Lebecque S, Dezutter-Dambuyant C, Schmitt D, Zlotnik A et al (2000) Macrophage inflammatory protein 3alpha is expressed at inflamed epithelial surfaces and is the most potent chemokine known in attracting Langerhans cell precursors. *J Exp Med* 192: 705–718
98 Liao F, Rabin RL, Smith CS, Sharma G, Nutman TB, Farber JM (1999) CC-chemokine receptor 6 is expressed on diverse memory subsets of T cells and determines responsiveness to macrophage inflammatory protein 3 alpha. *J Immunol* 162: 186–194
99 Dieu MC, Vanbervliet B, Vicari A, Bridon JM, Oldham E, Ait-Yahia S, Briere F, Zlotnik A, Lebecque S, Caux C (1998) Selective recruitment of immature and mature dendritic

cells by distinct chemokines expressed in different anatomic sites. *J Exp Med* 188: 373–386
100 Campbell DJ, Butcher EC (2002) Rapid acquisition of tissue-specific homing phenotypes by CD4(+) T cells activated in cutaneous or mucosal lymphoid tissues. *J Exp Med* 195: 135–141
101 Mora JR, Bono MR, Manjunath N, Weninger W, Cavanagh LL, Rosemblatt M, Von Andrian UH (2003) Selective imprinting of gut-homing T cells by Peyer's patch dendritic cells. *Nature* 424: 88–93
102 Stagg AJ, Kamm MA, Knight SC (2002) Intestinal dendritic cells increase T cell expression of alpha4beta7 integrin. *Eur J Immunol* 32: 1445–1454
103 Dudda JC, Martin SF (2004) Tissue targeting of T cells by DCs and microenvironments. *Trends Immunol* 25: 417–421
104 Mora JR, Cheng G, Picarella D, Briskin M, Buchanan N, von Andrian UH (2005) Reciprocal and dynamic control of CD8 T cell homing by dendritic cells from skin- and gut-associated lymphoid tissues. *J Exp Med* 201: 303–316. Epub 10 January 2005
105 Iwata M, Hirakiyama A, Eshima Y, Kagechika H, Kato C, Song SY (2004) Retinoic acid imprints gut-homing specificity on T cells. *Immunity* 21: 527–538
106 Dudda JC, Simon JC, Martin S (2004) Dendritic cell immunization route determines $CD8^+$ T cell trafficking to inflamed skin: role for tissue microenvironment and dendritic cells in establishment of T cell-homing subsets. *J Immunol* 172: 857–863
107 Kakinuma T, Nakamura K, Wakugawa M, Mitsui H, Tada Y, Saeki H, Torii H, Komine M, Asahina A, Tamaki K (2002) Serum macrophage-derived chemokine (MDC) levels are closely related with the disease activity of atopic dermatitis. *Clin Exp Immunol* 127: 270–273
108 Kakinuma T, Saeki H, Tsunemi Y, Fujita H, Asano N, Mitsui H, Tada Y, Wakugawa M, Watanabe T, Torii H et al (2003) Increased serum cutaneous T cell-attracting chemokine (CCL27) levels in patients with atopic dermatitis and psoriasis vulgaris. *J Allergy Clin Immunol* 111: 592–597
109 Flier J, Boorsma DM, van Beek PJ, Nieboer C, Stoof TJ, Willemze R, Tensen CP (2001) Differential expression of CXCR3 targeting chemokines CXCL10, CXCL9, and CXCL11 in different types of skin inflammation. *J Pathol* 194: 398–405

Chemokine biology of NK cells and γδ T cells

Chenggang Jin and Craig T. Morita

Division of Rheumatology, Department of Internal Medicine and the Interdisciplinary Group in Immunology, University of Iowa College of Medicine, EMRB 340F, Iowa City, IA 52242, USA

Introduction

Natural killer (NK) cells and γδ T cells are populations of lymphocytes that mediate immunity against pathogens and malignant tumors. Both generally exhibit significant cytotoxicity, produce high levels of inflammatory cytokines and chemokines, and share the expression of cellular receptors (generally designated NK receptors) for the detection of MHC class I and class Ib proteins and other cell surface proteins.

NK cells are large granular lymphocytes that play important roles in host defense against viral, bacterial, and parasitic infections as well as in the surveillance for malignant cells. In addition to their cytotoxic capabilities, NK cells serve as regulators of immune responses through the release of a variety of cytokines and chemokines. NK cells are bone marrow-derived lymphocytes that were originally characterized by their ability to spontaneously mediate lysis of certain tumor cell lines, their large granular morphology, and their lack of a T cell receptor and CD3 complex. NK cells do not use the specialized gene rearrangement machinery that assembles T and B cell antigen receptors. Instead, NK cells express both inhibitory and activating cell surface receptors. Inhibitory receptors include C-type lectin family receptors, such as Ly49 and CD94/NKG2, or Ig superfamily receptors such as killer immunoglobulin-like receptors (KIR). These receptors generally recognize MHC class I (class Ia) and class Ib (HLA-E in human and Qa-1 in mice) proteins. A number of activating receptors on NK cells have been described that are alternative forms of Ly49, KIR family (termed KAR), and CD94/NKG2 receptors and that have similar specificities as their inhibitory forms. In addition, unique activating receptors are also expressed such as the NKG2D receptor that recognizes MICA/B, Rae1, and H60, CD16 that binds to IgG, NKp44/NKp46 that recognize viral hemagglutinins, and NKp30 whose ligands are not well characterized. The primary peripheral NK cells present in humans and mice are mature cells with decreasing frequencies in blood, spleen, and bone marrow, respectively. Human NK cells comprise 15% of all circulating lymphocytes and can be divided into two subsets, $CD56^{bright}$ and $CD56^{dim}$, each subset having unique functional attributes and distinct roles in the human immune response [1].

Chemokine Biology – Basic Research and Clinical Application, Volume I
edited by Bernhard Moser, Gordon L. Letts and Kuldeep Neote

γδ T cells function as a bridge between the innate and adaptive immune systems by killing infected and malignant cells and by functioning as a source of cytokines and chemokines involved in activation of immune cells and in maintaining tissue integrity [2]. Although γδ T cells express a rearranged γδ T cell receptor and exhibit memory responses to nonpeptide antigens [3], they also share properties with NK cells including the expression of inhibitory and activating NK receptors and other NK markers. These unique T cells constitute a small proportion (5%) of T cells in the peripheral blood and lymphoid organs of human and rodents whereas they are the major population of T cells in ruminants and chickens [3]. γδ T cells are enriched in epithelial tissue such as the skin (in mice and other species but not humans), the intestine (in most species), and the reproductive tract.

Unlike αβ T cells, γδ T cells, at certain anatomical sites, display a highly restricted usage of Vγ and Vδ genes. For example, in mice, epithelial γδ T cells that constitute the vast majority of skin T cells all express a single invariant T cell antigen receptor [4]. In humans, γδ T cells generally use one of three major Vδ gene segments, Vδ1, Vδ2, or Vδ3, and the majority of adult human peripheral γδ T cells express Vγ2 paired with Vδ2 [3]. Human Vγ2Vδ2 T cells mediate both immediate effector functions and memory responses by using their T cell antigen receptor (TCR) to recognize nonpeptide phosphorylated intermediates found in isoprenoid and other metabolic pathways of microbes and humans [3]. Human Vδ1-bearing γδ T cells recognize lipids presented by CD1 [5] and the MHC class I chain-related genes MICA and MICB (major histocompatibility complex class I chain-related antigen A and B) [6]. Antigens for murine γδ T cells include a stress-induced compound produced by keratinocytes [7] and the TL MHC class Ib molecule [8]. In addition to stimulation through the γδ T cell antigen receptor, γδ T cells can also be activated or inhibited by the NK receptors that they commonly express [9–11].

Unlike NK cells, γδ T cells have the ability to mount memory responses and to differentiate into T cells that retain memory. Thus, in mice, many γδ T cells acquire a memory phenotype and behave similarly to memory αβ T cells [12]. Similarly, many human γδ T cells have memory T cell phenotypes [13–15] that can be acquired even before birth (our unpublished observations). Studies on BCG infection in monkeys [16], and BCG vaccination in humans [17], provide evidence for memory responses by Vγ2Vδ2 T cells.

NK cells

Chemokines produced by NK cells

NK cells serve as effectors and regulators of immune responses through direct cytotoxicity and through the release of a variety of cytokines and chemokines including

IFN-γ, TNF-α, GM-CSF, IL-5, CXCL8, CCL3, CCL4, and CCL5 [18, 19]. NK cells can produce a number of these chemokines without specific activation. Unstimulated human peripheral blood NK cells can produce CCL3, CCL5, and CCL22 *in vitro* [20]. Isolated human NK cells also express mRNA for XCL1 (lymphotactin) [21]. $CD56^{bright}CD16^{-}$ NK cells isolated from human early pregnancy deciduas express CXCL8 mRNA and secrete large amounts of CXCL8 [22]. Moreover, purified peripheral blood NK cells, derived from elderly healthy subjects older than 90, produced CXCL8, CCL3, and CCL5 [23].

Chemokine production by NK cells can be induced or further increased when NK cells are activated *in vitro* by soluble factors or *in vivo* by infection. NK cells can be activated by IL-2 or IL-12 resulting in the increased synthesis of CXCL8, CCL3, and CCL5 [23]. NK cells also produce CCL22 upon stimulation [24]. Immunohistochemical analysis of IL-2-activated murine NK cells and Northern analysis of human NK clones revealed that these cells also produce XCL1, a chemokine that attracts both NK and T cells *in vivo* [25]. Large amounts of CXCL8 were produced when purified NK cells were stimulated with IL-18 (IFN-γ-inducing factor) [26]. Ligation of β1 integrins on human NK cells also results in the production of CXCL8, through the activation of the Rac1/p38 mitogen-activated protein kinase (MAPK) signaling pathway [27]. Treatment of human and mouse NK cells with ULBP (human cytomegalovirus glycoprotein UL16 binding protein) ligands for the activating receptor, NKG2D/DAP10, led to increased production of IFN-γ, TNF-α, and CCL4 [28]. Soluble ULBP1, -2, and -3 fusion proteins stimulated production of GM-CSF, TNF-β, and CCL1. Combining IL-12 and soluble ULBP2 had a strong synergistic effect on CCL1 production [29]. Thus, activated NK cells can produce CXCL8, XCL1, CCL1, CCL3, CCL4, CCL5, and CCL22.

The production of chemokines by NK cells plays an important role in their function. CCL3 along with IFN-γ are required for protective NK cell responses *in vivo* to murine cytomegalovirus (MCMV) infection [30]. $CCL3^{-/-}$ mice have decreased resistance to MCMV due, at least in part, to dramatically reduced NK cell accumulations as well as decreased IFN-γ production in their livers [30, 31]. These data suggest that NK cell production of IFN-γ and chemokines may be coordinated to control MCMV and that NK cells may be triggered through different mechanisms during their response to infections.

NK cell activation through the NK receptor, Ly49H, or through cytokines such as IL-2, IL-12, IL-15, IL-18, and type I IFN can stimulate NK cells to produce IFN-γ, XCL1, CCL3, CCL4, and CCL5. Both mechanisms contribute to the *in vivo* response of NK cells [32, 33]. During infection with *Listeria*, IFN-γ, XCL1, CCL3, CCL4, and CCL5 are coexpressed at the single-cell level in activated NK cells, $CD8^{+}$ T cells, and $CD4^{+}$ Th1 cells [34, 35]. In MCMV infection, murine NK cells produce these five mediators either after triggering of Ly49H or after exposure to innate cytokines. Cross-linking the activating Ly49D mouse NK receptor *in vitro* also induces high levels of IFN-γ, XCL1, CCL3, and CCL4 [36]. The local release of the five cytokines/chemokines by $Ly49H^+$ NK cells probably attracts and activates neighboring target cells, such as macrophages and dendritic cells, as well as other NK cells. In HIV-infected patients, NK cells were also shown to produce high levels of the CC-chemokines, CCL3, CCL4, and CCL5, which suppress HIV-1 entry and replication *in vitro* [20, 37, 38]. These findings suggest that NK cells play essential roles in recruiting inflammatory effector cells during infection and have the capacity to organize and shape adaptive immune responses.

Expression of chemokine receptors on NK cells

Different chemokine receptors are expressed on the two major human NK cell subsets that are identified by expression of CD56 and CD16 ($CD56^{dim}CD16^+$ and $CD56^{bright}CD16^-$) [39, 40]. Resting $CD56^{dim}CD16^+$ NK cells uniformly express CXCR1, CXCR4, and CX3CR1 at high levels. CXCR2 and CXCR3 were present at lower levels. There was no detectable surface expression of CC chemokine receptors (CCR1–7, 9) or CXCR5 or CXCR6 (Bonzo) using available antibodies [1]. As expected, resting $CD56^{dim}CD16^+$ NK cells migrated vigorously to CXCL12 and CX3CL1. The expression of CXCR1, CXCR2, and CX3CR1 on NK cells was also confirmed by other groups [41, 42]. However, Inngjerdingen et al. found that purified, resting human NK cells expressed CXCR4 but not CXCR1, CXCR2, CXCR3, or CX3CR1 [21]. One possible explanation for this discrepancy is that some chemokine receptors may be downregulated during the purification of NK cells [21].

In contrast to $CD56^{dim}CD16^+$ cells, resting $CD56^{bright}CD16^-$ NK cells express little CXCR1 or CX3CR1 but high levels of CCR5, CCR7, CXCR3, and CXCR4 [1]. $CD56^{bright}CD16^-$ NK cells exhibited chemotaxis to CCL5 [1], CCL19 [43], CCL21 [43], CXCL10 [1], CXCL11 [1], CCL22 [44], and CXCL12 [43].

Expression of CC chemokine receptors on activated NK cells has also been reported and is similar to $CD56^{bright}CD16^-$ NK cells. Whereas resting $CD56^{dim}CD16^+$ NK cells do not express CC chemokine receptors, activated human NK cells express CCR2, CCR4, CCR5, and CCR8 [45, 46]. Murine NK cells have been shown to express CCR2 and exhibit *in vitro* chemotaxis to the CCR2 ligand, CCL2 [45]. Studies using antibody neutralization and CCR2 gene knockout ($CCR2^{-/-}$) mice demonstrated that early recruitment of NK cells to the lungs is critically dependent on CCL2 and that disruption of this early recruitment results in increased severity of infection [47]. As determined by flow cytometric, immunoblot, and RNase protection assays, Inngjerdingen et al. showed that IL-2-activated human NK cells express CCR4 and CCR8 and respond to CCL17, CCL22, and CCL1 [46]. Cytolytic activity of NK cells is also augmented by CCL2, CCL3, CCL4, and CCL5 [48]. Moreover, proliferation of $CD56^{dim}CD16^+$ NK cells is costimulated by CCL19 and CCL 21 [21].

Expression of CCR7 on a subset of human NK cells was reported by several groups. CCR7 is an important determinant for T cell homing to secondary lymphoid organs through high endothelial venules. Earlier work showed that mRNA for CCR7 was not detected in resting peripheral blood NK cells. Consistent with this fact, CCL19 and CCL21 did not induce chemotaxis of resting NK cells. However, adult and cord blood NK cell population isolated by positive selection using CD56 beads showed strong chemotactic activity for NK cells to CCL19 and CCL21 [43]. Campbell et al. showed that $CD16^-$ NK cells express CCR7 and respond to the CCR7 ligands, CCL19 and CCL21 [1]. Consistent with this result, $CD56^{bright}CD16^-$ NK cells have been found in peripheral lymph nodes [49, 50]. Besides CCR7, CXCR3 may also mediate NK localization to lymph nodes. CXCR3, rather than CCR7, mediates recruitment of murine NK cells to lymph nodes that are undergoing an immune response [51]. Importantly, NK cells at stimulated lymph nodes provide an initial source of IFN-γ that is necessary for T_H1 polarization [51]. Expression of CCR7, therefore, differs between different NK subsets and CCR7 may be downregulated in resting NK cells.

γδ T cells

γδ T cells produce a number of different chemokines

γδ T cells not only express a range of chemokine receptors, but also produce chemokines. Therefore, γδ T cells can mediate their effector functions directly through cell killing or indirectly by cytokine production or by recruiting or regulating other cells. Human γδ T cells are commonly polarized to a Th1 phenotype and produce large amount of proinflammatory chemokines, such as CCL3, CCL4, CCL5, and XCL1, but not CCL2 or the Th2 chemoattractants, CCL1 and CCL17 [52–54]. Some Th2-polarized γδ T cells exist and have been found to secrete CXCL8, CCL1, and CCL17 [52]. Cipriani et al. showed that activation of human peripheral blood $V\delta2^+$ cells with the nonpeptide antigen, isopentenyl pyrophosphate (IPP), rapidly stimulates the production of the C-C chemokines, CCL3 and CCL4, and XCL1 but not CCL2. IPP stimulation of Vγ2Vδ2 production of CCL3 and CCL4 was not affected by IL-4, IL-10, TGF-β, or IFN-γ. However, IL-12 significantly enhanced IPP-induced expression and release of CCL3. CCL3 release was downregulated by TGF-β whereas the induction of CCL4 by IPP and IL-12 was refractory to inhibition by TGF-β [53]. Upon IPP stimulation, peripheral blood γδ T cells also increase the production of CXCL8. Amplification of CXCL8 expression may be increased by the interaction between the activation marker CD30 on γδ T cells, and its ligand CD30L constitutively expressed by neutrophils [55].

In murine models of bacterial infection, γδ T cells have been reported to participate in host defense against extracellular bacteria such as *E. coli* and intracellular bacteria such as *Listeria monocytogenes* [56–58]. Tagawa et al. found that the number of murine γδ T cells bearing invariant Vγ6Vδ1 significantly increased in the peritoneal cavity during a peritoneal infection with *E. coli* [59]. To elucidate potential roles of invariant Vδ1-bearing γδ T cells in protection against *E. coli* infection, Tagawa et al. examined bacterial growth and cellular responses in the peritoneal cavities of mice deficient in Vδ1 (Vδ1$^{-/-}$) following peritoneal infection with *E. coli.* Vδ1$^{-/-}$ mice showed severely impaired accumulation of peritoneal macrophages after *E. coli* infection. The peritoneal γδ T cells of infected wild-type mice produced large amounts of chemokines such as CCL3 and CCL5 in response to γδ TCR triggering *in vitro*, whereas there was no production of those cytokines by peritoneal γδ T cells of Vδ1$^{-/-}$ mice. Thus, Vδ1^{+} γδ T cells may help to augment innate immunity by secreting chemokines that attract macrophages [59].

Chemokine secretion is likely to contribute to the effector functions of γδ T cells. Human Vγ2Vδ2 T cells have been shown to suppress HIV-1 replication *in vitro* via the production of CCL3, CCL4, and CCL5 chemokines that bind the HIV-1 coreceptor, CCR5, preventing HIV entry [60]. In the SIV macaque model, immunization with SIVgp120 and p27 protected the animals from subsequent challenge with live SIV by the rectal mucosal route. In the protected macaques, γδ T cells in the rectal mucosa were increased and found to produce the CCR5 ligands, CCL3, CCL4, and CCL5 [61]. The production of the same chemokines by γδ T cells might worsen autoimmune diseases such as multiple sclerosis or its animal model, experimental allergic encephalomyelitis (EAE) [62–64].

Intraepithelial γδ T cells are not only attracted to epithelial surfaces by chemokines but also produce chemokines that recruit other cells. Boismenu et al. showed that murine Vγ3Vδ1 (also termed Vγ5Vδ1) dendritic epidermal T cells (DETC) can produce CCL3, CCL4, CCL5, and XCL1 but not CCL2 [65]. XCL1 mRNA was also detected in stimulated γδ intraepithelial lymphocytes (IEL) isolated from the small intestine of these mice [65]. Expression of XCL1, CCL3, CCL4, and CCL5 mRNA was specifically detected in intestinal γδ IEL on gene microarray analysis whereas these chemokines were absent (CCL3 and CCL4) or much less abundant (CCL5 and XCL1) in αβ IEL [66]. Consistent with these findings, murine γδ T cells were required for production of the CXCL1, XCL1, CCL3, and CCL4 chemokines in response to thermal injury in the gut but not in the lung [67]. Freshly isolated and activated human intestinal intraepithelial γδ T cells also expressed high levels of CXCL8 mRNA [68].

CCR9/CCL25 and CCR10/CCL27direct γδ T cells to the small intestine and the skin

Chemokines and chemokine receptors have been shown to play key roles in determining tissue-specific homing of hematopoietic cells [69, 70]. The unique expression pattern of CCR9 and the distribution of its ligand, CCL25, suggest that CCR9 and CCL25 play important roles in thymocyte development and lymphocyte migration to the gut. CCR9 is an excellent example of an organ-specific chemokine receptor, because its ligand, CCL25, is selectively expressed in the small intestine and thymus [71]. CCR9 has been demonstrated to be the chemokine receptor that regulates lymphocyte trafficking during T cell development and in the gut [72]. The thymus has been shown to express various chemokines, including XCL1, CCL3, CCL4, CCL17, CCL19, CCL21, CCL25, and CXCL12 [73]. This profile of chemokine expression in thymus appears to suggest that chemokines may play an important role in thymopoiesis. Expression of CCL25 was detected in medullary dendritic cells, thymic epithelial cells, and small intestine epithelial cells. CCL25 may be important for the development, homeostasis, and/or function of mucosal αβ and γδ T cells. CCR9 is expressed on the majority of immature $CD4^+CD8^+$ (double positive, DP) thymocytes, and is downregulated during their maturation into the $CD4^+$ or $CD8^+$ stage. These findings suggest that CCR9 may be involved in regulating T cell migration within the thymus. Half of murine γδ thymocytes and peripheral γδ T cells express CCR9 [74], suggesting CCR9 and its chemokine, CCL25, may function in the development and trafficking of γδ T cells. In bone marrow transplantation experiments, $CCR9^{-/-}$ bone marrow cells showed a reduced capacity to repopulate the thymus compared with bone marrow cells from $CCR9^{+/+}$ mice [74, 75]. Studies with $CCR9^{-/-}$ mice also showed that CCR9 expression is required for the migration of γδ T cells to the small intestine. $CCR9^{-/-}$ mice had increased number of peripheral γδ T cells but reduced number of γδ intraepithelial lymphocytes (IEL) in the small intestine suggesting that without CCR9, γδ T cells do not migrate to the gut. Thus, CCR9/CCL25 plays an important role in regulating the development and migration of γδ T cells [71].

Similar to the gut, γδ T cells also constitute the primary T cell population in murine skin. Skin γδ T cells have a dendritic appearance that maximizes their contact with keratinocytes and are termed dendritic epidermal T cells (DETC). CCR10 is expressed by a subset of skin-homing-memory T cells including DETC and αβ T cells [76, 77]. Some tissue cells also express CCR10 mRNA but the function of CCR10 in these cells is not known. Two CCR10 ligands, CCL27 and CCL28, have been identified [78]. CCL27 is selectively and constitutively produced by keratinocytes. CCL27 can mediate the preferential migration of skin-homing, $CCR10^+$, CLA-bearing memory T cells *in vitro*. In mice, expression of the CCR10 chemokine receptor by Vγ3Vδ1 DETC was dramatically upregulated in the $CD122^+$ (IL-2Rβ^+) population from both wild-type (Vγ3^+) and Vγ2^+ transgenic mice [79]. Using *in vitro* chemotaxis assays, Xiong et al. showed that Vγ3^+ $CD122^+$ cells preferentially migrated towards CCL27, demonstrating that CCR10 on γδ T cells is functional. Stimulation of γδ T cells though their T-cell receptors (TCRs) significantly upregulated CCR10 expression by Vδ4 and Vδ5 $CD122^-$ T cells. Although CCR10 expression on skin γδ T cells has not been directly tested in humans (where γδ T cells represent 1–4% of skin T cells), most CD3⁺ T cells in affected skin from patients with atopic dermatitis or psoriasis expressed CCR10 [80, 81]. Thus, engagement of TCRs on murine γδ T cells may induce upregulation of the CCR10 chemokine receptor allowing their homing to the epidermis. Selective expression of CCR10 on activated fetal thymic γδ T cells may direct their preferential migration to adult and fetal skin in response to CCL27.

Distinct chemokine receptor profile on human γδ T cell subsets

Glatzel et al. [82] and our laboratory in collaboration with B. Moser (unpublished data [83, 84]) showed that human γδ T cells express a variety of chemokine receptors, including CXC and CC receptors. We found significant differences between γδ and αβ T cells in the expression of all chemokine receptors analyzed except CCR3, CCR4, and CCR6. Strikingly, a significant proportion of γδ T cells expressed the innate or acute phase chemokine receptors, CXCR1 and CXCR2, which are prominently expressed on neutrophils. These chemokine receptors are typically expressed by innate immune cells such as neutrophils, monocytes, and NK cells and in inflammed tissue, but not by conventional αβ T cells [85]. Moreover, unlike most αβ T cells, some γδ T cells also express CCR1 and CCR2. In addition to the expression of CXCR1, CXCR2, and CCR1, a higher proportion of γδ T cells expressed chemokine receptors responding to inducible chemokines that are characteristic of Th1/Tc1αβ T cells including CXCR3, CXCR6, and CCR5. Glatzel et al. also found that a high proportion of γδ T cells express CCR5 and CXCR3 chemokine receptors, whereas a lower proportion of αβ T cells express CCR5 and CXCR3. Only a small proportion of γδ T cells expressed CCR4 and the other two Th2-associated chemokine receptors, CCR3 and CCR8, were not detected on γδ T cells. This expression pattern of chemokine receptors is consistent with the cytokine profile of γδ T cells since they produce large amount of IFN-γ and TNF-α after stimulated with non-peptide antigens. Almost no γδ T cells express the CXCL13/BCA-1 receptor CXCR5 that defines a subset of CD4⁺ αβ T cells that home to B cell follicles. Also, only a fraction of γδ T cells express the lymph node homing receptor CCR7 compared with a high proportion of CD4 or CD8 αβ T cells. These results clearly demonstrate that γδ T cells express a distinct array of chemokine receptors when compared with αβ T cells, favoring innate (CXCR1 and CXCR2) and Th1/Tc1 (CXCR3, CXCR6, and CCR5) but not lymphoid tissue-homing receptors (CXCR5 and CCR7).

Chemokine receptor expression by neonatal γδ T cells was also studied in our laboratory. Unlike neonatal αβ T cells, some neonatal γδ T cells express CXCR5, CXCR6, CCR6, and CCR9 although like αβ T cells they also expressed CXCR3, CXCR4, and CCR7. As noted, a higher proportion of adult γδ T cells expressed CXCR1, CXCR2, CXCR3, CXCR6, CCR1, CCR2, and CCR5 compared with αβ T cells and neonatal γδ T cells.

Chemokine receptors on human Vδ1 and Vδ2 T cells

γδ T cells expressing Vδ1 and Vδ2 TCR comprise the majority of peripheral blood γδ T cells. Vδ2+ T cells predominate in most adults due to an environmentally dependent expansion of γδ T cells in infancy [86]. Vδ2 T cells are the dominant γδ T cells in circulating blood, whereas Vδ1 T cells are preferentially found in such tissues as gut epithelium and skin. When the two major V gene subsets of human γδ T cells were examined, Vδ1 and Vδ2 T cells were found to have different surface markers and recognize different classes of antigens [13]. We compared the chemokine receptor expression by the two subsets of γδ T cells. Although there were no differences in CXCR1 and CXCR2 expression between Vδ1 and Vδ2 T cells, few Vδ1 T cells expressed inflammatory CXCR6, CCR1, CCR2, CCR5, and CCR6 that are expressed by Vγ2Vδ2 T cells (unpublished observations). Although the proportion of cells was lower, similar differences were also noted in neonates (unpublished observations). These findings are consistent with Vδ1 T cells being primarily late or effector memory cells that express CD45RA and lack CD62L [13].

Memory subsets of human Vγ2Vδ2 T cells express different arrays of chemokine receptors

Human Vγ2Vδ2 T cells can be divided into distinct subsets according to different surface markers, proliferative ability, and effector functions [87–89]. Expression of CD27, CD28, and CD45RO [90] or CCR7 [91] have been used to distinguish the different memory subsets of T cells. Dieli et al. [15] and our laboratory have examined chemokine receptor expression by different memory Vγ2Vδ2 T cell subsets (Tab. 1). Strikingly, we find that a high proportion of intermediate ($CD27^+CD28^-CD45RA/RO^+$) and late ($CD27^-CD28^-CD45RA^+$) memory Vγ2Vδ2 T cells expressed CXCR1 and CXCR2 chemokine receptors and CD56 and did not express CCR6 or CCR7. In contrast, most early ($CD27^{+/-}CD28^+CD45RO^+$) memory Vγ2Vδ2 T cells selectively expressed CXCR6, CCR1, and CCR2 while only a minor fraction expressed CCR6 and CCR7. A high proportion of both subsets expressed CXCR3, and CCR5 chemokine receptors characteristic of Th1/Tc1 cells [83, 92, 93] and CXCR4 [83]. A higher proportion of early memory γδ T cells than early memory CD8 αβ T cells expressed CXCR3, CXCR6, CCR1, CCR2, and CCR5. Also, there was no difference in the expression of other chemokine receptors when $CCR7^+$ and $CCR7^-$ early memory Vγ2Vδ2 T cells were compared (unpublished observations).

To determine if CXCR1 receptors expressed by intermediate and late memory γδ T cells were functional, we measured chemotaxis of purified blood γδ, NK, and αβ T cells to CXCL8 (a ligand for CXCR1) and CXCL12 (a ligand for CXCR4). Consistent with the observed expression profiles, γδ T and NK cells specifically migrat-

Table 1 - Expression of chemokine receptors by subsets of human NK cells and Vγ2Vδ2 T cells[a]

Chemokine receptor	NK cells[b]		Vγ2Vδ2 T cells[c]			
	$CD56^{bright}CD16^-$	$CD56^{dim}CD16^+$	T_{Early}[d]	$T_{Early\ 27-}$	T_{Int}	$T_{Late\ RA}$
CX3CR1	–[e]	++[f]	–/+[g]	–/+	++	++
CXCR1	–	++	–/+	–/+	++	++
CXCR2	–	++	–/+	–/+	++	++
CXCR3	+[h]	–/+	++	++	++	+
CXCR4	+	+	+	+	+	+
CXCR5	–	–	–	–	–	–
CXCR6	–	–	++	++	–/+	–/+
CCR1	–	–	++	++	+	+
CCR2	–	–	++	++	–/+	–/+
CCR3	–	–	–	–	–	–
CCR4	–	–/+	–/+	–/+	–/+	–/+
CCR5	+	–	++	++	++	++
CCR6	–	–	+	–/+	–	–/+
CCR7	+	–	+	–/+	–	–/+
CCR8	ND[i]	ND	–	–	–	–
CCR9	–	–	–	–	–	–
CCR10	ND	ND	ND	ND	ND	ND
XCR1	ND	ND	ND	ND	ND	ND

[a]*Data are based on surface molecule expression detected by FACS*
[b]*References [1, 21]*
[c]*References (Our unpublished data and [15, 82, 83, 84])*
[d]*T_{Early}, $CD28^+CD27^+$; $T_{Early\ 27-}$, $CD28^+CD27^-$; T_{Int}, $CD28^-CD27^+$; $T_{Late\ RA}$, $CD27^-CD28^-CD45RA^+$*
[e]*Not expressed*
[f]*High-density expression by the majority of cells*
[g]*Low-density and variable expression by the minority of cells*
[h]*Expression by about half of the cells*
[i]*No data available*

ed to CXCL8 whereas αβ T cells did not. All three populations migrated to CXCL12. γδ T cells expressing CXCR1 also fluxed calcium and had increased actin polymerization with exposure to CXCL8. γδ T cells were also able to migrate to the CXCR3 ligand, CXCL10, and the CCR1/3/5 ligand, CCL5. Consistent with their proposed effector function, $CXCR1^+$ intermediate and late memory Vγ2Vδ2 T cells also expressed more perforin than early memory Vγ2Vδ2 T cells. In contrast, early memory and intermediate memory cells showed stronger proliferation to nonpeptide antigens than CD45RA$^+$ late memory Vγ2Vδ2 T cells.

We also found that a high proportion of human γδ T cells expressed the β_7 intergrin chain that directs T cells to epithelial surfaces or to Peyer's patches when paired with α_E or α_4, respectively. Also some γδ T cells expressed the cutaneous lymphocyte antigen (CLA) that is required for skin homing. β_7 and CLA expression did not affect chemokine receptor expression since γδ T cells that expressed or did not express these molecules had identical proportions of cells expressing the different chemokine receptors. Given the high proportion of resting γδ T cells expressing CLA and β_7 receptors as well as E- and P-selectin ligands [94], a high proportion of resting human γδ T cells are able to migrate to epithelial and other peripheral tissues for antigen recognition and effector functions but not to lymph nodes.

A role for γδ T cells in humoral immunity

As detailed above, human peripheral blood γδ T cells mainly express receptors for Th1/Tc1 inflammatory chemokines but not lymph node chemokines. Brandes et al. found that freshly isolated Vγ2Vδ2 T cells expressed CXCR3 and CCR5, as expected, but almost no CCR7 [83]. Accordingly, these cells migrated and fluxed calcium when exposed to CXCL11 and CCL5, but not CCL21 [83]. Stimulation with the nonpeptide antigen, IPP, fundamentally changed the migration properties of Vγ2Vδ2 T cells by rapidly inducing CCR7, and to a lesser extent CCR4, and dramatically down-modulating CCR5 and to a lesser extent CCR2 [83]. Maximal expression of CCR7 was reached early (12–36 h) after Vγ2Vδ2 T cell activation and declined to baseline after 2 weeks, indicating that CCR7 was primarily expressed early after activation. Similar upregulation of CCR7 has been reported for memory αβ T cells *in vitro* [95]. This inverse relationship in the regulation of inflammatory versus lymph node homing chemokine receptors (e.g., CCR5 *versus* CCR7) after γδ T cell stimulation paralleled changes in the migratory responses to the corresponding chemokines, CCL5 and CCL21.

Thus far, CCR7 has been linked with the relocation of lymphocytes to secondary lymphoid tissues, including lymph nodes and Peyer's patches. In agreement, γδ T cells have been detected within lymph nodes. Immunohistochemical analysis directly showed that γδ T cells clustered within the follicular dendritic cell network of germinal centers in lymph nodes [83]. The localization in germinal centers implies a role for γδ T cells in humoral immune responses. In support of such a role and consistent with earlier studies [96–99], co-culture of B cells with γδ T cells results in the production of substantial levels of IgM, IgG, and IgA [83]. These findings suggest the involvement of activated γδ T cells in humoral immunity during antimicrobial responses.

Conclusions

NK and γδ T cells have both similarity and difference in chemokine production and chemokine receptor expression. Both cell types produce CCL3, CCL4, CCL5, and CXCL8, and express the innate chemokine receptors, CX3CR1, CXCR1, and CXCR2, unlike most αβ T cells. However, like Th1/Tc1 αβ T cells, γδ T cells express Th1/Tc1 chemokine receptors such as CXCR3, CXCR6, CCR2, and CCR5, whereas NK cells do not express CXCR6, CCR1, and CCR2. Subsets of NK cells and memory subsets of γδ T cells exhibit distinct migratory potentials by expressing different chemokine receptors. $CD56^{dim}CD16^{+}$ NK cells and intermediate and late memory Vγ2Vδ2 T cells express high level of CX3CR1, CXCR1, and CXCR2, whereas $CD56^{bright}CD16^{-}$ NK cells and early memory Vγ2Vδ2 T cells generally do not express these chemokine receptors (Tab. 1). The role of chemokines and chemokine receptors expressed by NK cells and γδ T cells in both immunopathogenesis and immune functions are not well understood. While chemokines produced by NK and γδ T cells suppress HIV-1 entry and replication *in vitro*, these same chemokines are also involved in the pathogenesis of multiple sclerosis. Furthermore, high level expression of the major HIV coreceptor CCR5, rendered γδ T cells more vulnerable to HIV infection *in vitro*. Further studies are needed to clarify the roles of chemokines and chemokine receptors in NK and γδ T cell functions. Such knowledge will pave the way for the development of new therapies based on blocking or stimulating interactions between chemokines and their receptors.

Acknowledgements
We thank Hong Wang, Ghanashyam Sarikonda, and Amy M. Raker for critical reading of the manuscript. Supported by NIH NIAMS grant AR-045504 (to C.T.M.).

References

1 Campbell JJ, Qin S, Unutmaz D, Soler D, Murphy KE, Hodge MR, Wu L, Butcher EC (2001) Unique subpopulations of $CD56^{+}$ NK and NK-T peripheral blood lymphocytes identified by chemokine receptor expression repertoire. *J Immunol* 166: 6477–6482
2 Morita CT, Mariuzza RA, Brenner MB (2000) Antigen recognition by human γδ T cells: pattern recognition by the adaptive immune system. *Springer Semin Immunopathol* 22: 191–217
3 Morita CT, Mariuzza RA, Brenner MB (2000) Antigen recognition by human γδ T

cells: pattern recognition by the adaptive immune system. *Springer Semin Immunopathol* 22: 191–218
4 Hayday A, Tigelaar R (2003) Immunoregulation in the tissues by γδ T cells. *Nat Rev Immunol* 3: 233–242
5 Spada FM, Grant EP, Peters PJ, Sugita M, Melián A, Leslie DS, Lee HK, van Donselaar E, Hanson DA, Krensky AM et al. (2000) Self recognition of CD1 by γδ T cells: Implications for innate immunity. *J Exp Med* 191: 937–948
6 Groh V, Steinle A, Bauer S, Spies T (1998) Recognition of stress-induced MHC molecules by intestinal epithelial γδ T cells. *Science* 279: 1737–1740
7 Havran WL, Chien Y-h, Allison JP (1991) Recognition of self antigens by skin-derived T cells with invariant γδ receptors. *Science* 252: 1430–1432
8 Crowley MP, Faher AM, Baumgarth N, Hampl J, Gutgemann I, Teyton L, Chien Y-h (2000) A population of murine γδ T cells recognized an inducible MHC class Ib molecule. *Science* 287: 314–316
9 Carena I, Shamshiev A, Donda A, Colonna M, De Libero G (1997) Major histocompatibility complex class I molecules modulate activation threshold and early signaling of T cell antigen receptor-γ/δ stimulated by nonpeptidic ligands. *J Exp Med* 186: 1769–1774
10 Halary F, Peyrat M-A, Champagne E, Lopez-Botet M, Moretta A, Moretta L, Vié H, Fournié J-J, Bonneville M (1997) Control of self-reactive cytotoxic T lymphocytes expressing γδ T cell receptors by natural killer inhibitory receptors. *Eur J Immunol* 27: 2812–2821
11 Poccia F, Cipriani B, Vendetti S, Colizzi V, Poquet Y, Battistini L, López-Botet M, Fournié JJ, Gougeon M-L (1997) CD94/NKG2 inhibitory receptor complex modulates both anti-viral and anti-tumoral responses of polyclonal phosphoantigen-reactive Vγ9Vδ2 T lymphocytes. *J Immunol* 159: 6009–6017
12 Tough DF, Sprent J (1998) Lifespan of γ/δ T cells. *J Exp Med* 187: 357–365
13 De Rosa SC, Mitra DK, Watanabe N, Herzenberg LA, Roederer M (2001) Vd1 and Vd2 γδ T cells express distinct surface markers and might be developmentally distinct lineages. *J Leukoc Biol* 70: 518–526
14 De Rosa SC, Andrus JP, Perfetto SP, Mantovani JJ, Herzenberg LA, Roederer M (2004) Ontogeny of γδ T cells in humans. *J Immunol* 172: 1637–1645
15 Dieli F, Poccia F, Lipp M, Sireci G, Caccamo N, Di Sano C, Salerno A (2003) Differentiation of effector/memory Vδ2 T cells and migratory routes in lymph nodes or inflammatory sites. *J Exp Med* 198: 391–397
16 Shen Y, Zhou D, Qiu L, Lai X, Simon M, Shen L, Kou Z, Wang Q, Jiang L, Estep J et al. (2002) Adaptive immune response of Vγ2Vδ2^{+} T cells during mycobacterial infections. *Science* 295: 2255–2258
17 Hoft DF, Brown RM, Roodman ST (1998) Bacille Calmette-Guérin vaccination enhances human γδ T cell responsiveness to mycobacteria suggestive of a memory-like phenotype. *J Immunol* 161: 1045–1054
18 Biron CA, Nguyen KB, Pien GC, Cousens LP, Salazar-Mather TP (1999) Natural

killer cells in antiviral defense: function and regulation by innate cytokines. *Annu Rev Immunol* 17: 189–220

19 Raulet DH, Vance RE, McMahon CW (2001) Regulation of the natural killer cell receptor repertoire. *Annu Rev Immunol* 19: 291–330

20 Oliva A, Kinter AL, Vaccarezza M, Rubbert A, Catanzaro A, Moir S, Monaco J, Ehler L, Mizell S, Jackson R et al. (1998) Natural killer cells from human immunodeficiency virus (HIV)-infected individuals are an important source of CC-chemokines and suppress HIV-1 entry and replication *in vitro*. *J Clin Invest* 102: 223–231

21 Inngjerdingen M, Damaj B, Maghazachi AA (2001) Expression and regulation of chemokine receptors in human natural killer cells. *Blood* 97: 367–375

22 Saito S, Kasahara T, Sakakura S, Enomoto M, Umekage H, Harada N, Morii T, Nishikawa K, Narita N, Ichijo M (1994) Interleukin-8 production by CD16-$CD56^{bright}$ natural killer cells in the human early pregnancy decidua. *Biochem Biophys Res Commun* 200: 378–383

23 Mariani E, Meneghetti A, Neri S, Ravaglia G, Forti P, Cattini L, Facchini A (2002) Chemokine production by natural killer cells from nonagenarians. *Eur J Immunol* 32: 1524–1529

24 Andrew DP, Chang MS, McNinch J, Wathen ST, Rihanek M, Tseng J, Spellberg JP, Elias CG 3rd (1998) STCP-1 (MDC) CC chemokine acts specifically on chronically activated Th2 lymphocytes and is produced by monocytes on stimulation with Th2 cytokines IL-4 and IL-13. *J Immunol* 161: 5027–5038

25 Hedrick JA, Saylor V, Figueroa D, Mizoue L, Xu Y, Menon S, Abrams J, Handel T, Zlotnik A (1997) Lymphotactin is produced by NK cells and attracts both NK cells and T cells *in vivo*. *J Immunol* 158: 1533–1540

26 Puren AJ, Fantuzzi G, Gu Y, Su MS, Dinarello CA (1998) Interleukin-18 (IFNγ-inducing factor) induces IL-8 and IL-1β via TNFα production from non-$CD14^+$ human blood mononuclear cells. *J Clin Invest* 101: 711–721

27 Mainiero F, Soriani A, Strippoli R, Jacobelli J, Gismondi A, Piccoli M, Frati L, Santoni A (2000) RAC1/P38 MAPK signaling pathway controls β1 integrin-induced interleukin-8 production in human natural killer cells. *Immunity* 12: 7–16

28 Kubin M, Cassiano L, Chalupny J, Chin W, Cosman D, Fanslow W, Mullberg J, Rousseau AM, Ulrich D, Armitage R (2001) ULBP1, 2, 3: novel MHC class I-related molecules that bind to human cytomegalovirus glycoprotein UL16, activate NK cells. *Eur J Immunol* 31: 1428–1437

29 Cosman D, Mullberg J, Sutherland CL, Chin W, Armitage R, Fanslow W, Kubin M, Chalupny NJ (2001) ULBPs, novel MHC class I-related molecules, bind to CMV glycoprotein UL16 and stimulate NK cytotoxicity through the NKG2D receptor. *Immunity* 14: 123–133

30 Salazar-Mather TP, Orange JS, Biron CA (1998) Early murine cytomegalovirus (MCMV) infection induces liver natural killer (NK) cell inflammation and protec-

tion through macrophage inflammatory protein 1α (MIP-1α)-dependent pathways. *J Exp Med* 187: 1–14
31 Salazar-Mather TP, Hamilton TA, Biron CA (2000) A chemokine-to-cytokine-to-chemokine cascade critical in antiviral defense. *J Clin Invest* 105: 985–993
32 Hodge DL, Schill WB, Wang JM, Blanca I, Reynolds DA, Ortaldo JR, Young HA (2002) IL-2 and IL-12 alter NK cell responsiveness to IFN-γ-inducible protein 10 by down-regulating CXCR3 expression. *J Immunol* 168: 6090–6098
33 Bluman EM, Bartynski KJ, Avalos BR, Caligiuri MA (1996) Human natural killer cells produce abundant macrophage inflammatory protein-1α in response to monocyte-derived cytokines. *J Clin Invest* 97: 2722–2727
34 Dorner BG, Scheffold A, Rolph MS, Huser MB, Kaufmann SH, Radbruch A, Flesch IE, Kroczek RA (2002) MIP-1α, MIP-1β, RANTES, and ATAC/lymphotactin function together with IFN-γ as type 1 cytokines. *Proc Natl Acad Sci USA* 99: 6181–6186
35 Dorner BG, Smith HR, French AR, Kim S, Poursine-Laurent J, Beckman DL, Pingel JT, Kroczek RA, Yokoyama WM (2004) Coordinate expression of cytokines and chemokines by NK cells during murine cytomegalovirus infection. *J Immunol* 172: 3119–3131
36 Ortaldo JR, Bere EW, Hodge D, Young HA (2001) Activating Ly-49 NK receptors: central role in cytokine and chemokine production. *J Immunol* 166: 4994–4999
37 Fehniger TA, Herbein G, Yu H, Para MI, Bernstein ZP, O'Brien WA, Caligiuri MA (1998) Natural killer cells from HIV-1$^+$ patients produce C-C chemokines and inhibit HIV-1 infection. *J Immunol* 161: 6433–6438
38 Simmons G, Reeves JD, Hibbitts S, Stine JT, Gray PW, Proudfoot AE, Clapham PR (2000) Co-receptor use by HIV and inhibition of HIV infection by chemokine receptor ligands. *Immunol Rev* 177: 112–126
39 Cooper MA, Fehniger TA, Caligiuri MA (2001) The biology of human natural killer-cell subsets. *Trends Immunol* 22: 633–640
40 Lanier LL, Le AM, Civin CI, Loken MR, Phillips JH (1986) The relationship of CD16 (Leu-11) and Leu-19 (NKH-1) antigen expression on human peripheral blood NK cells and cytotoxic T lymphocytes. *J Immunol* 136: 4480–4486
41 Chuntharapai A, Lee J, Hébert CA, Kim KJ (1994) Monoclonal antibodies detect different distribution patterns of IL-8 receptor A and IL-8 receptor B on human peripheral blood leukocytes. *J Immunol* 153: 5682–5688
42 Morohashi H, Miyawaki T, Nomura H, Kuno K, Murakami S, Matsushima K, Mukaida N (1995) Expression of both types of human interleukin-8 receptors on mature neutrophils, monocytes, and natural killer cells. *J Leukoc Biol* 57: 180–187
43 Kim CH, Pelus LM, Appelbaum E, Johanson K, Anzai N, Broxmeyer HE (1999) CCR7 ligands, SLC/6Ckine/Exodus2/TCA4 and CKβ-11/MIP-3β/ELC, are chemoattractants for CD56$^+$CD16$^-$ NK cells and late stage lymphoid progenitors. *Cell Immunol* 193: 226–235
44 Godiska R, Chantry D, Raport CJ, Sozzani S, Allavena P, Leviten D, Mantovani A,

Gray PW (1997) Human macrophage-derived chemokine (MDC), a novel chemoattractant for monocytes, monocyte-derived dendritic cells, and natural killer cells. *J Exp Med* 185: 1595–1604

45 Polentarutti N, Allavena P, Bianchi G, Giardina G, Basile A, Sozzani S, Mantovani A, Introna M (1997) IL-2-regulated expression of the monocyte chemotactic protein-1 receptor (CCR2) in human NK cells: characterization of a predominant 3.4-kilobase transcript containing CCR2B and CCR2A sequences. *J Immunol* 158: 2689–2694

46 Inngjerdingen M, Damaj B, Maghazachi AA (2000) Human NK cells express CC chemokine receptors 4 and 8 and respond to thymus and activation-regulated chemokine, macrophage-derived chemokine, and I-309. *J Immunol* 164: 4048–4054

47 Morrison BE, Park SJ, Mooney JM, Mehrad B (2003) Chemokine-mediated recruitment of NK cells is a critical host defense mechanism in invasive aspergillosis. *J Clin Invest* 112: 1862–1870

48 Maghazachi AA, Al-Aoukaty A, Schall TJ (1996) CC chemokines induce the generation of killer cells from $CD56^+$ cells. *Eur J Immunol* 26: 315–319

49 Fehniger TA, Cooper MA, Nuovo GJ, Cella M, Facchetti F, Colonna M, Caligiuri MA (2003) $CD56^{bright}$ natural killer cells are present in human lymph nodes and are activated by T cell-derived IL-2: a potential new link between adaptive and innate immunity. *Blood* 101: 3052–3057

50 Ferlazzo G, Thomas D, Lin SL, Goodman K, Morandi B, Muller WA, Moretta A, Munz C (2004) The abundant NK cells in human secondary lymphoid tissues require activation to express killer cell Ig-like receptors and become cytolytic. *J Immunol* 172: 1455–1462

51 Martín-Fontecha A, Thomsen LL, Brett S, Gerard C, Lipp M, Lanzavecchia A, Sallusto F (2004) Induced recruitment of NK cells to lymph nodes provides IFN-γ for T_H1 priming. *Nat Immunol* 5: 1260–1265

52 Dagna L, Iellem A, Biswas P, Resta D, Tantardini F, Fortis C, Sabbadini MG, D'Ambrosio D, Manfredi AA, Ferrarini M (2002) Skewing of cytotoxic activity and chemokine production, but not of chemokine receptor expression, in human type-1/-2 γδ T lymphocytes. *Eur J Immunol* 32: 2934–2943

53 Cipriani B, Borsellino G, Poccia F, Placido R, Tramonti D, Bach S, Battistini L, Brosnan CF (2000) Activation of C-C β-chemokines in human peripheral blood γδ T cells by isopentenyl pyrophosphate and regulation by cytokines. *Blood* 95: 39–47

54 Poccia F, Battistini L, Cipriani B, Mancino G, Martini F, Gougeon ML, Colizzi V (1999) Phosphoantigen-reactive Vγ9Vδ2 T lymphocytes suppress *in vitro* human immunodeficiency virus type 1 replication by cell-released antiviral factors including CC chemokines. *J Infect Dis* 180: 858–861

55 Biswas P, Rovere P, De Filippi C, Heltai S, Smith C, Dagna L, Poli G, Manfredi AA, Ferrarini M (2000) Engagement of CD30 shapes the secretion of cytokines by human γδ T cells. *Eur J Immunol* 30: 2172–2180

56 Nakamura T, Matsuzaki G, Nomoto K (1999) The protective role of T-cell receptor

Vγ1⁺ T cells in primary infection with *Listeria monocytogenes*. *Immunology* 96: 29–34
57 Matsuzaki G, Yamada H, Kishihara K, Yoshikai Y, Nomoto K (2002) Mechanism of murine Vg1⁺ γδ T cell-mediated innate immune response against *Listeria monocytogenes* infection. *Eur J Immunol* 32: 928–935
58 Skeen MJ, Freeman MM, Ziegler HK (2004) Changes in peritoneal myeloid populations and their proinflammatory cytokine expression during infection with Listeria monocytogenes are altered in the absence of γ/δ T cells. *J Leukoc Biol* 76: 104–115
59 Tagawa T, Nishimura H, Yajima T, Hara H, Kishihara K, Matsuzaki G, Yoshino I, Maehara Y, Yoshikai Y (2004) Vδ1⁺ γδ T cells producing CC chemokines may bridge a gap between neutrophils and macrophages in innate immunity during *Escherichia coli* infection in mice. *J Immunol* 173: 5156–5164
60 Poggi A, Carosio R, Fenoglio D, Brenci S, Murdaca G, Setti M, Indiveri F, Scabini S, Ferrero E, Zocchi MR (2004) Migration of Vδ1 and Vδ2 T cells in response to CXCR3 and CXCR4 ligands in healthy donors and HIV-1-infected patients: competition by HIV-1 Tat. *Blood* 103: 2205–2213
61 Lehner T, Mitchell E, Bergmeier L, Singh M, Spallek R, Cranage M, Hall G, Dennis M, Villinger F, Wang Y (2000) The role of γδ T cells in generating antiviral factors and β-chemokines in protection against mucosal simian immunodeficiency virus infection. *Eur J Immunol* 30: 2245–2256
62 Rajan AJ, Klein JD, Brosnan CF (1998) The effect of γδ T cell depletion on cytokine gene expression in experimental allergic encephalomyelitis. *J Immunol* 160: 5955–5962
63 Rajan AJ, Asensio VC, Campbell IL, Brosnan CF (2000) Experimental autoimmune encephalomyelitis on the SJL mouse: effect of γδ T cell depletion on chemokine and chemokine receptor expression in the central nervous system. *J Immunol* 164: 2120–2130
64 Murzenok PP, Matusevicius D, Freedman MS (2002) γ/δ T cells in multiple sclerosis: chemokine and chemokine receptor expression. *Clin Immunol* 103: 309–316
65 Boismenu R, Feng L, Xia YY, Chang JCC, Havran WL (1996) Chemokine expression by intraepithelial γδ T cells. Implications for the recruitment of inflammatory cells to damaged epithelia. *J Immunol* 157: 985–992
66 Fahrer AM, Konigshofer Y, Kerr EM, Ghandour G, Mack DH, Davis MM, Chien Y-H (2001) Attributes of γδ intraepithelial lymphocytes as suggested by their transcriptional profile. *Proc Natl Acad Sci USA* 98: 10261–10266
67 Toth B, Alexander M, Daniel T, Chaudry IH, Hubbard WJ, Schwacha MG (2004) The role of γδ T cells in the regulation of neutrophil-mediated tissue damage after thermal injury. *J Leukoc Biol* 76: 545–552
68 Lundqvist C, Melgar S, Yeung MM, Hammarström S, Hammarström ML (1996) Intraepithelial lymphocytes in human gut have lytic potential and a cytokine profile that suggest T helper 1 and cytotoxic functions. *J Immunol* 157: 1926–1934

69 Moser B, Loetscher P (2001) Lymphocyte traffic control by chemokines. *Nat Immunol* 2: 123–128
70 Moser B, Wolf M, Walz A, Loetscher P (2004) Chemokines: multiple levels of leukocyte migration control. *Trends Immunol* 25: 75–84
71 Wurbel MA, Malissen M, Guy-Grand D, Meffre E, Nussenzweig MC, Richelme M, Carrier A, Malissen B (2001) Mice lacking the CCR9 CC-chemokine receptor show a mild impairment of early T- and B-cell development and a reduction in T-cell receptor $\gamma\delta^+$ gut intraepithelial lymphocytes. *Blood* 98: 2626–2632
72 Zabel BA, Agace WW, Campbell JJ, Heath HM, Parent D, Roberts AI, Ebert EC, Kassam N, Qin S, Zovko M et al. (1999) Human G protein-coupled receptor GPR-9-6/CC chemokine receptor 9 is selectively expressed on intestinal homing T lymphocytes, mucosal lymphocytes, and thymocytes and is required for thymus-expressed chemokine-mediated chemotaxis. *J Exp Med* 190: 1241–1256
73 Papadakis KA, Landers C, Prehn J, Kouroumalis EA, Moreno ST, Gutierrez-Ramos JC, Hodge MR, Targan SR (2003) CC chemokine receptor 9 expression defines a subset of peripheral blood lymphocytes with mucosal T cell phenotype and Th1 or T-regulatory 1 cytokine profile. *J Immunol* 171: 159–165
74 Uehara S, Song K, Farber JM, Love PE (2002) Characterization of CCR9 expression and CCL25/thymus-expressed chemokine responsiveness during T cell development: $CD3^{high}CD69^+$ thymocytes and $\gamma\delta$ TCR^+ thymocytes preferentially respond to CCL25. *J Immunol* 168: 134–142
75 Uehara S, Grinberg A, Farber JM, Love PE (2002) A role for CCR9 in T lymphocyte development and migration. *J Immunol* 168: 2811–2819
76 Jarmin DI, Rits M, Bota D, Gerard NP, Graham GJ, Clark-Lewis I, Gerard C (2000) Identification of the orphan receptor G-protein-coupled receptor 2 as CCR10, a specific receptor for the chemokine ESkine. *J Immunol* 164: 3460–3464
77 Homey B, Wang W, Soto H, Buchanan ME, Wiesenborn A, Catron D, Muller A, McClanahan TK, Dieu-Nosjean MC, Orozco R et al. (2000) The orphan chemokine receptor G protein-coupled receptor-2 (GPR-2, CCR10) binds the skin-associated chemokine CCL27 (CTACK/ALP/ILC). *J Immunol* 164: 3465–3470
78 Wang W, Soto H, Oldham ER, Buchanan ME, Homey B, Catron D, Jenkins N, Copeland NG, Gilbert DJ, Nguyen N et al. (2000) Identification of a novel chemokine (CCL28), which binds CCR10 (GPR2). *J Biol Chem* 275: 22313–22323
79 Xiong N, Kang C, Raulet DH (2004) Positive selection of dendritic epidermal $\gamma\delta$ T cell precursors in the fetal thymus determines expression of skin-homing receptors. *Immunity* 21: 121–131
80 Reiss Y, Proudfoot AE, Power CA, Campbell JJ, Butcher EC (2001) CC chemokine receptor (CCR)4 and the CCR10 ligand cutaneous T cell-attracting chemokine (CTACK) in lymphocyte trafficking to inflamed skin. *J Exp Med* 194: 1541–1547
81 Homey B, Alenius H, Muller A, Soto H, Bowman EP, Yuan W, McEvoy L, Lauerma AI, Assmann T, Bunemann E et al. (2002) CCL27-CCR10 interactions regulate T cell-mediated skin inflammation. *Nat Med* 8: 157–165

82 Glatzel A, Wesch D, Schiemann F, Brandt E, Janssen O, Kabelitz D (2002) Patterns of chemokine receptor expression on peripheral blood γδ T lymphocytes: strong expression of CCR5 is a selective feature of Vδ2/Vγ9 γδ T cells. *J Immunol* 168: 4920–4929

83 Brandes M, Willimann K, Lang AB, Nam K-H, Jin C, Brenner MB, Morita CT, Moser B (2003) Flexible migration program regulates γδ T-cell involvement in humoral immunity. *Blood* 102: 3693–3701

84 Roth SJ, Diacovo TG, Brenner MB, Rosat J-P, Buccola J, Morita CT, Springer TA (1998) Transendothelial chemotaxis of human α/β and γ/δ T lymphocytes to chemokines. *Eur J Immunol* 28: 104–113

85 Williams EJ, Haque S, Banks C, Johnson P, Sarsfield P, Sheron N (2000) Distribution of the interleukin-8 receptors, CXCR1 and CXCR2, in inflamed gut tissue. *J Pathol* 192: 533–539

86 Parker CM, Groh V, Band H, Porcelli SA, Morita C, Fabbi M, Glass D, Strominger JL, Brenner MB (1990) Evidence for extrathymic changes in the T cell receptor γ/δ repertoire. *J Exp Med* 171: 1597–1612

87 Gioia C, Agrati C, Casetti R, Cairo C, Borsellino G, Battistini L, Mancino G, Goletti D, Colizzi V, Pucillo LP et al. (2002) Lack of CD27-CD45RA-Vγ9Vδ2^{+} T cell effectors in immunocompromised hosts and during active pulmonary tuberculosis. *J Immunol* 168: 1484–1489

88 De Rosa SC, Andrus JP, Perfetto SP, Mantovani JJ, Herzenberg LA, Roederer M (2004) Ontogeny of γδ T cells in humans. *J Immunol* 172: 1637–1645

89 Eberl M, Engel R, Beck E, Jomaa H (2002) Differentiation of human γ-δ T cells towards distinct memory phenotypes. *Cell Immunol* 218: 1–6

90 Appay V, Rowland-Jones SL (2004) Lessons from the study of T-cell differentiation in persistent human virus infection. *Semin Immunol* 16: 205–212

91 Sallusto F, Geginat J, Lanzavecchia A (2004) Central memory and effector memory T cell subsets: function, generation, and maintenance. *Annu Rev Immunol* 22: 745–763

92 Kim CH, Rott L, Kunkel EJ, Genovese MC, Andrew DP, Wu L, Butcher EC (2001) Rules of chemokine receptor association with T cell polarization *in vivo*. *J Clin Invest* 108: 1331–1339

93 Sallusto F, Lenig D, Mackay CR, Lanzavecchia A (1998) Flexible programs of chemokine receptor expression on human polarized T helper 1 and 2 lymphocytes. *J Exp Med* 187: 875–883

94 Diacovo TG, Roth SJ, Morita CT, Rosat J-P, Brenner MB, Springer TA (1996) Interactions of human α/β and γ/δ T lymphocyte subsets in shear flow with E-selectin and P-selectin. *J Exp Med* 183: 1193–1203

95 Sallusto F, Kremmer E, Palermo B, Hoy A, Ponath P, Qin S, Forster R, Lipp M, Lanzavecchia A (1999) Switch in chemokine receptor expression upon TCR stimulation reveals novel homing potential for recently activated T cells. *Eur J Immunol* 29: 2037–2045

96 Rajagopalan S, Zordan T, Tsokos GC, Datta SK (1990) Pathogenic anti-DNA autoantibody-inducing T helper cell lines from patients with active lupus nephritis: isolation of CD4⁻8⁻ T helper cell lines that express the γδ T-cell antigen receptor. *Proc Natl Acad Sci USA* 87: 7020–7024
97 van Vlasselaer P, Gascan H, de Waal Malefyt R, de Vries JE (1992) IL-2 and a contact-mediated signal provided by TCRαβ$^+$ or TCRγδ$^+$CD4$^+$ T cells induce polyclonal Ig production by committed human B cells. Enhancement by IL-5, specific inhibition of IgA synthesis by IL-4. *J Immunol* 148: 1674–1684
98 Munk ME, Fazioli RA, Calich VL, Kaufmann SHE (1995) Paracoccidioides brasiliensis-stimulated human γ/δ T cells support antibody production by B cells. *Infect Immun* 63: 1608–1610
99 Horner AA, Jabara H, Ramesh N, Geha RS (1995) γ/δ T lymphocytes express CD40 ligand and induce isotype switching in B lymphocytes. *J Exp Med* 181: 1239–1244

Dendritic cell traffic control by chemokines

Federica Sallusto, Alfonso Martín-Fontecha and Antonio Lanzavecchia

Institute for Research in Biomedicine, Via Vincenzo Vela 6, CH-6500 Bellinzona, Switzerland

Introduction

Dendritic cells (DCs) are widely accepted as the most potent and versatile antigen-presenting cells. They have an extraordinary capacity to acquire and process antigens for presentation to T cells and to express high levels of major histocompatibility complex (MHC) molecules and co-stimulatory molecules that drive naïve T cell activation. In addition DCs produce cytokines, primarily IL-12, which contribute to shape the quality of the T cell response generated. The capacity to migrate to sites of inflammation and from there to the T cell areas of secondary lymphoid organs is a fundamental aspect of DC biology. It has become apparent that the large families of chemokines and chemokine receptors provide a flexible code for regulating DC traffic and positioning in both homeostatic and inflammatory conditions.

Dissemination of DC precursors and immature DCs under steady state and inflammatory conditions

Under steady state conditions immature DCs seed into all bodily tissues where they reside as "sentinels" ready to react to incoming pathogens, a state that is defined as immature [1]. Langerhans cells (LCs) are a subset of immature DCs resident in epithelia and characterised by a relatively slow turnover [2]. LCs contain characteristic endosomal structures, called Birbeck granules, organised by a LC-specific lectin (Langerin) and are anchored to epithelial cells through E-cadherin. LCs express CCR6, the receptor for CCL20, a chemokine which is produced constitutively by keratinocytes [3]. Human monocytes cultured in the presence of TGF-β acquire some of the cardinal features of LCs, such as expression of Langerin [4], raising the possibility that LCs differentiate from peripheral monocytes under the aegis of local cytokines.

Chemokine Biology – Basic Research and Clinical Application, Volume I
edited by Bernhard Moser, Gordon L. Letts and Kuldeep Neote

Immature DCs are also present in the dermis and in all parenchyma. These cells have a turnover of approximately 2–4 days and need to be continuously replenished by precursors derived from the blood [2, 5]. The precursors of tissue DCs have not been fully characterised. They may be circulating immature DCs [6], which are present in low numbers in peripheral blood, or monocytes. The latter represent an abundant source of DC precursors that can be recruited at sites of inflammation or infection where they rapidly differentiate to DCs [7, 8]. Monocytes express CCR2 that promotes extravasation into inflamed tissues and migration towards a gradient of inflammatory chemokines. Mice deficient in CCR2 or in its ligand MCP-1 have impaired immune responses that appear to be due to defective monocyte migration both in the afferent and efferent phase of the immune response [9, 10]. Monocytes also express CCR5, a receptor for inflammatory chemokines and a co-receptor for HIV, and CXCR4, the receptor for CXCL12. It is possible that CXCR4 may be involved in the constitutive traffic of monocytes and DCs into certain tissues including tumours where hypoxia induces high levels of CXCL12 production.

In mice peripheral blood monocytes are a heterogeneous population comprising at least two functional subsets: a short-lived CX_3CR1^{lo} $CCR2^+$ $Gr1^+$ subset that is actively recruited to inflamed tissues and a CX_3CR1^{hi} $CCR2^-$ Gr1- subset characterised by CX_3CR1-dependent recruitment to non-inflamed tissues [11]. Both subsets have the potential to differentiate into DCs *in vivo*. The level of CX_3CR1 expression also defines two major human monocyte subsets, the $CD14^+$ $CD16^-$ and $CD14^{lo}$ $CD16^+$ monocytes, which share phenotype and homing potential with the mouse subsets [12]. Recently a subset of circulating monocytes, identified as $Gr1^{int}$, has been identified that selectively expresses CCR7 and CCR8 [13]. These monocytes may be disposed to become lymphatic-migrating DCs. When these monocyte-derived DCs exit skin to emigrate to lymph nodes, they may use not only CCR7, as it will be described below, but also CCR8.

A distinct subset of DCs, called plasmacytoid DCs (pDCs) or interferon-producing cells (IPCs) has been described in humans [14–16] and, more recently, in mouse [17]. Although IPCs are capable of presenting endogenous antigen to $CD8^+$ T lymphocytes [18], their hallmark function is the production of high amounts of type I IFN following viral infection. Immature IPC precursors circulating in peripheral blood express CXCR3 and CXCR4 as well as L-selectin and E/P-selectin ligands (PSGL-1 and CLA). This pattern of expression would be consistent with the capacity of these cells to migrate both to inflamed lymph nodes and peripheral tissues where CXCR3 and CXCR4 ligands are displayed on endothelial cells. Indeed, IPCs are typically localised around high endothelial venules (HEV) in inflamed lymph nodes and in some inflamed tissues [19–21]. Migration of IPCs require the coordinate action of CXCR3 and CXCR4, possibly through a mechanism that entails features of haptotaxis, i.e., dependency on chemokine immobilisation, and chemorepulsion, i.e., movement away from highest chemokine concentration [21, 22].

DC maturation: effects on chemokine receptor expression and chemokine production

The DC maturation process can be induced by a variety of stimuli. The most effective are microbial products that trigger specific Toll-like receptors (TLRs) on DCs. Interestingly human myeloids DCs (mDCs) and IPCs express complementary sets of TLRs and consequently respond to different agonists [23, 24]. In particular, mDCs express TLR1, TLR2, TLR3, TLR4, TLR5, TLR6, and TLR8 whereas IPCs express TLR7 and TLR9. Thus, while risiquimod (a synthetic compound that triggers in humans TLR7 and TLR8) triggers both DC types, LPS (a TLR4 agonist) selectively triggers mDCs and CpG (a TLR9 agonist) selectively triggers IPCs. In addition, DC maturation can be induced by inflammatory cytokines, such as IL-1 and TNF, as well as by endogenous "danger signals" released by necrotic cells, such as heat shock proteins and urate crystals [25]. CD40L is also a potent DC maturation stimulus but since it is delivered by activated T cells it acts primarily as a secondary stimulus that enhances cytokine production initially elicited by microbial stimulation [26, 27].

Using a global gene expression approach it has been recently shown that the maturation program induced by TLR triggering involves the coordinate regulation of approximately 8,000 genes that control several DC functions ranging from antigen capture and presentation to co-stimulation, cytokine production and chemokine expression and responsiveness [28]. While most of the genes appear to be triggered by almost all stimuli a few genes have a high activation threshold. Indeed genes involved in the differentiation of Th1 and inflammatory T cells, such as IL-12, IL-23 and Delta-4, have been found to be elicited only in response to combinations of selected TLR ligands which act in synergy [28].

In response to microbial products DCs produce high amounts of inflammatory chemokines, up to the extraordinary amount of 2 pg/cell of CCL4 [29]. These chemokines, which include CCL2, CCL4 and CCL5, are produced very rapidly but only for a limited period of time and may play two distinct functions: first they attract DC precursors at sites of antigen exposure; second, by inducing a rapid and complete internalisation of the cognate receptors on maturing DCs allow these cells to exit the tissue. Indeed, CCR1 and CCR5 disappear within 1 h from the surface of maturing DCs while they remain detectable intracellularly for several days [29]. Eventually, however, these receptors are downregulated at the mRNA level. At later time points following induction of maturation DCs express CCL17 and CCL19 that attract CCR4 and CCR7 positive cells, respectively, and may thus favour interaction with naïve and activated T cells [30].

A common feature of maturing DCs and IPCs is the upregulation of CCR7, the receptor for CCL19 and CCL21. CCL21 is constitutively expressed in lymphatic endothelial cells and high endothelial venules and is involved in the recruitment of maturing DCs and other $CCR7^+$ cells at these sites [31]. CCL21 is expressed together with CCL19 by stromal cells in the T cell areas in a lymphotoxin β-dependent fashion. CCL19 is also produced by maturing DCs at late time points after stimulation and is therefore expected to be released primarily in the lymph node. CCR7 expression and responsiveness gradually increased in maturing DCs. This receptor also shows a striking resistance to ligand-induced downregulation, indicating that DCs can sustain the response to CCL19 and CCL21 throughout the maturation process. The transcriptional regulation of the CCR7 gene has not been characterised. In general CCR7 expression is induced by stimuli that induce upregulation of MHC and co-stimulatory molecules. However, there are examples of maturation stimuli that do not induce CCR7 expression and stimuli that induce CCR7 expression independently of maturation. An example of the latter is the uptake of apoptotic cells by human monocyte-derived DCs that induces CCR7 expression and DC chemotaxis in response to CCL21, but results in downregulation of HLA-DR and CD86 [32].

DC traffic from sites of antigen capture to sites of antigen presentation

Priming of naïve T cells requires the encounter with antigen-presenting DCs in the specialised T cell areas of secondary lymphoid organs (Fig. 1). In certain experimental conditions it has been shown that intact antigen present in peripheral tissues can be transported to lymph nodes through the lymph. There it can be captured and presented by lymph node resident DCs that, under steady state condition, represent an extensive network of poorly stimulatory cells still endowed with antigen capturing capacity [33, 34]. The major route of antigen delivery to the lymph node is represented by peripheral tissue-resident DCs that migrate to the draining lymph nodes. For instance, maturing DCs that have taken up antigen in the skin and have been stimulated by microbial products migrate into lymphatic vessels and localise to the T cell areas of the draining lymph node. Similarly, splenic immature DCs which are present in the marginal zone and are exposed to blood-borne antigens rapidly mature and migrate to the T cell area following intravenous injection of microbial products. Both these processes are dependent from CCR7 upregulation in mature DCs.

CCR7-deficient mice have a major defect in DC migration from tissue to lymph nodes and from the marginal zone to the T cell zone of spleen [35]. Adoptive transfer experiments formally demonstrated that CCR7-deficient DCs do not migrate when injected to normal CCR7-expressing hosts [36]. Two recent lines of evidence suggest that the CCR7-dependent pathway of migration can be boosted by inflammatory mediators. First, the lipid mediators cysteinyl leukotrienes and prostaglandin E2 enhance the sensitivity of CCR7 [37, 38]. Second, inflammatory cytokines such as TNF and IL-1 increase expression of CCL21 on lymphatic endothelial cells [36]. Both mechanisms enhance the entry of maturing DCs into lymphatic vessels and the migration to lymph nodes.

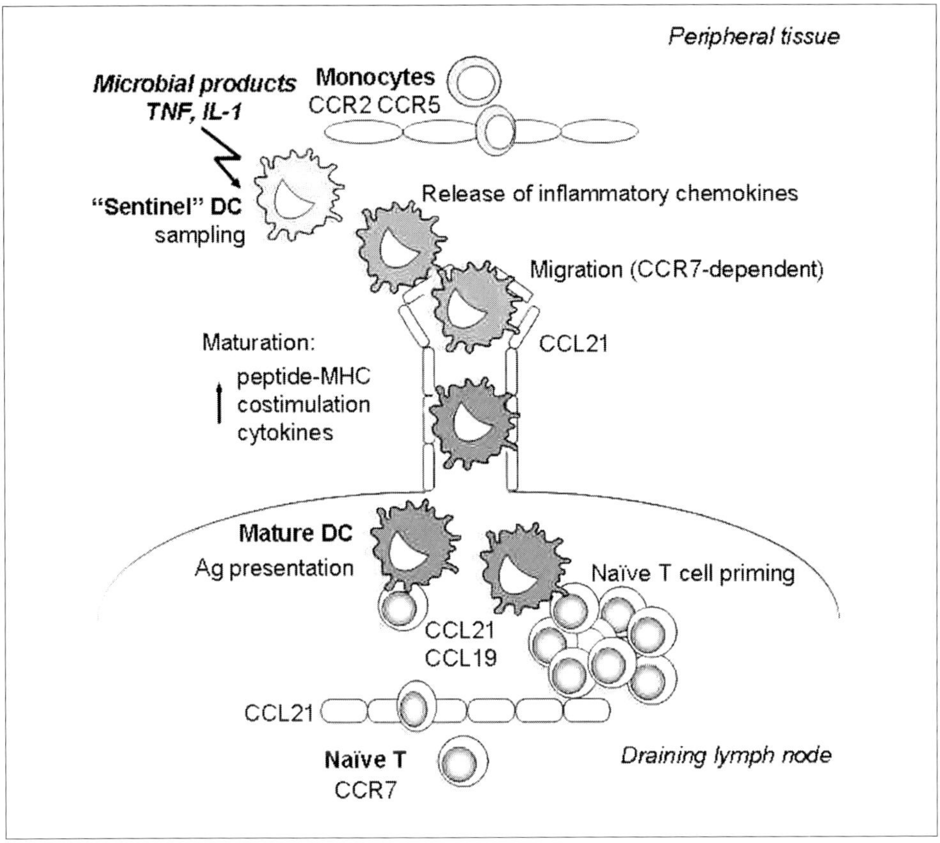

Figure 1
Immature "sentinel" DCs triggered by microbial products and inflammatory cytokines in peripheral tissues release inflammatory chemokines thus attracting DC precursors (monocytes) from the blood, and migrate in a CCR7-dependent fashion into lymphatic endothelial vessels. Maturing DCs upregulate MHC and co-stimulatory molecules and produce cytokines and chemokines, thus acquiring T cell priming and polarising capacity. Mature DCs localise in the T cell area where they present antigen to naïve T cells that home to the T cell area through a CCR7-dependent mechanism and induce their proliferation and differentiation to effector cells. Additional molecules, such as selectins and integrins, participate in these processes which are not depicted in the scheme.

Besides its role in driving the migration of antigen-carrying mature DCs in the course of an immune response, CCR7 appears to control the migration of DCs to lymph node in the steady state, a phenomenon that is much less understood. Mice lacking the adaptor molecule DAP12 present a homeostatic accumulation of DCs in peripheral sites, raising the possibility that a DAP12 linked receptor such as TREM-2 may play a role in controlling DC migration in homeostasis [39, 40].

Recent *in vivo* analysis using green fluorescent protein (GFP)-tagged cells revealed relevant differences between LCs and dermal DCs. After skin immunisation both LCs and dermal DCs migrate to the lymph node but the latter appear to migrate more rapidly, to colonise different areas, to express higher levels of co-stimulatory molecules and to be more capable of eliciting T cell responses [41]. Indeed, deletion of LCs did not impair the triggering of hapten-specific T cells.

DCs play an important role in the gut where they scan an enormous and continuously exposed surface. Mucosal DCs present in the lamina propria express CX_3CR1 which is required to form transepithelial dendrites, which enable DCs to directly sample luminal antigens, and commensal and pathogenic bacteria [42, 43]. These cells conditioned by local cytokines (for instance TGF-β) or T cells may regulate gut homeostasis, immunological tolerance and inflammation in the gut.

Impact of DC maturation and migration on T cell priming in physiological and vaccination settings

There is now abundant evidence that maturation state of antigen presenting DCs dictates the outcome of the T cell response. The most striking example is provided by the findings that in mice targeting of soluble antigens to lymph node resident immature DCs leads to an abortive T cell proliferation and establishment of tolerance whereas in the presence of a DC maturation stimulus, in the form of CD40 antibodies, the same antigen leads to effective T cell priming and generation of effector and memory cells [44].

In addition to the maturation state, the absolute number of antigen presenting DCs that migrate to the draining lymph node has a profound impact on the magnitude of the T cell response. This is particularly relevant in immunisation protocols in which antigen-loaded DCs are injected subcutaneously as cancer vaccines. In these protocols, human immature DCs are generated *in vitro* from haematopoietic progenitors or monocytes, pulsed with antigen in the forms of protein, peptide or mRNA, and induced to mature by stimulation with microbial products or inflammatory cytokines before injection [45]. In preclinical mouse systems subcutaneously injected mature DCs migrate to the lymph node in a CCR7-dependent fashion where they elicit T cell responses. In this setting the magnitude and quality of $CD4^+$ T cell response was proportional to the number of antigen-carrying DCs that reached the lymph node and could be boosted up to 40-fold by pre-injection of TNF that conditioned the tissue for increased DC migration by increasing the expression of the CCR7 ligand CCL21 in lymphatic endothelial cells [36]. Thus, lymphatic drainage of mature DCs can be manipulated to increase DC vaccine efficacy.

In mice mature DCs migrating to the draining lymph nodes rapidly recruit in a CCR7-independent, CXCR3-dependent manner natural killer (NK) cells, which are normally excluded from lymph nodes [46]. NK cell depletion and reconstitution experiments show that NK cells provide an early source of IFN-γ that is necessary for optimal Th1 polarisation. These results show that DCs can influence Th1 differentiation not only by elaborating Th1 promoting factors, such as IL-12, but also by recruiting to lymph node, through a yet undefined mechanism, NK cells that in some systems represent an essential source of IFN-γ for T cell polarisation.

Another factor that may influence T cell fate is the kinetics of DC activation. Recently migrated DCs actively produce Th1 polarising cytokines and effectively prime Th1 responses [47]. In contrast at late time points the same cells exhaust the IL-12 producing capacity and although still retaining T cell stimulatory capacity promote T cell proliferation without differentiation. Thus while "active" DCs induce differentiation of effector T cells, exhausted DCs may induce the development of memory T cells [48].

Conclusions

Gaining a better understanding of the migratory pathways of DCs in physiological settings will be essential for future advances in using DCs as a means to fine-tune immune responses in clinical settings such as in cancer, autoimmunity and transplantation. In the case of induction of anti-tumour response, strategies are being evaluated aiming at increasing the delivery of antigen-carrying mature DCs to lymph node to enhance the efficacy of the vaccine [49]. In other cases, such as in autoimmune disorders and transplantation, it may be beneficial to deliver to the lymph node immature tolerogenic DCs to dampen the immune response and induce and/or maintain peripheral tolerance. Interfering with the migration of DCs in the context of transplantation, i.e., blocking the reverse transmigration of donor DCs from the transplanted organ to the blood [50], is presently more difficult because the molecular mechanisms controlling this event are still poorly defined. Nonetheless also this approach holds promises as a yet another way to modulate the immune response by targeting DC migration.

References

1 Banchereau J, Steinman RM (1998) Dendritic cells and the control of immunity. *Nature* 392: 245–252

2 Merad M, Manz MG, Karsunky H, Wagers A, Peters W, Charo I, Weissman IL, Cyster JG, Engleman EG (2002) Langerhans cells renew in the skin throughout life under steady-state conditions. *Nat Immunol* 3: 1135–1141

3 Dieu-Nosjean MC, Massacrier C, Homey B, Vanbervliet B, Pin JJ, Vicari A, Lebecque S, Dezutter-Dambuyant C, Schmitt D, Zlotnik A et al (2000) Macrophage inflammatory protein 3alpha is expressed at inflamed epithelial surfaces and is the most potent chemokine known in attracting Langerhans cell precursors. *J Exp Med* 192: 705–718
4 Geissmann F, Prost C, Monnet JP, Dy M, Brousse N, Hermine O (1998) Transforming growth factor beta1, in the presence of granulocyte/macrophage colony-stimulating factor and interleukin 4, induces differentiation of human peripheral blood monocytes into dendritic Langerhans cells. *J Exp Med* 187: 961–966
5 Ruedl C, Koebel P, Bachmann M, Hess M, Karjalainen K (2000) Anatomical origin of dendritic cells determines their life span in peripheral lymph nodes. *J Immunol* 165: 4910–4916
6 Robert C, Fuhlbrigge RC, Kieffer JD, Ayehunie S, Hynes RO, Cheng G, Grabbe S, von Andrian UH, Kupper TS (1999) Interaction of dendritic cells with skin endothelium: A new perspective on immunosurveillance. *J Exp Med* 189: 627–636
7 Randolph GJ, Beaulieu S, Lebecque S, Steinman RM, Muller WA (1998) Differentiation of monocytes into dendritic cells in a model of transendothelial trafficking. *Science* 282: 480–483
8 Bruno L, Seidl T, Lanzavecchia A (2001) Mouse pre-immunocytes as non-proliferating multipotent precursors of macrophages, interferon-producing cells, CD8alpha(+) and CD8alpha(-) dendritic cells. *Eur J Immunol* 31: 3403–3412
9 Lu B, Rutledge BJ, Gu L, Fiorillo J, Lukacs NW, Kunkel SL, North R, Gerard C, Rollins BJ (1998) Abnormalities in monocyte recruitment and cytokine expression in monocyte chemoattractant protein 1-deficient mice. *J Exp Med* 187: 601–608
10 Kurihara T, Warr G, Loy J, Bravo R (1997) Defects in macrophage recruitment and host defense in mice lacking the CCR2 chemokine receptor. *J Exp Med* 186: 1757–1762
11 Geissmann F, Jung S, Littman DR (2003) Blood monocytes consist of two principal subsets with distinct migratory properties. *Immunity* 19: 71–82
12 Randolph GJ, Sanchez-Schmitz G, Liebman RM, Schakel K (2002) The CD16(+) (FcgammaRIII(+)) subset of human monocytes preferentially becomes migratory dendritic cells in a model tissue setting. *J Exp Med* 196: 517–527
13 Qu C, Edwards EW, Tacke F, Angeli V, Llodra J, Sanchez-Schmitz G, Garin A, Haque NS, Peters W, van Rooijen N et al (2004) Role of CCR8 and other chemokine pathways in the migration of monocyte derived dendritic cells to lymph nodes. *J Exp Med* 200: 1231–1241
14 Perussia B, Fanning V, Trinchieri G (1985) A leukocyte subset bearing HLA-DR antigens is responsible for *in vitro* alpha interferon production in response to viruses. *Nat Immun Cell Growth Regul* 4: 120–137
15 Grouard G, Rissoan MC, Filgueira L, Durand I, Banchereau J, Liu YJ (1997) The enigmatic plasmacytoid T cells develop into dendritic cells with interleukin (IL)-3 and CD40-ligand. *J Exp Med* 185: 1101–1111
16 Cella M, Jarrossay D, Facchetti F, Alebardi O, Nakajima H, Lanzavecchia A, Colonna

M (1999) Plasmacytoid monocytes migrate to inflamed lymph nodes and produce large amounts of type I interferon. *Nat Med* 5: 919–923

17 Asselin-Paturel C, Boonstra A, Dalod M, Durand I, Yessaad N, Dezutter-Dambuyant C, Vicari A, O'Garra A, Biron C, Briere F et al (2001) Mouse type I IFN-producing cells are immature APCs with plasmacytoid morphology. *Nat Immunol* 2: 1144–1150

18 Salio M, Palmowski MJ, Atzberger A, Hermans IF, Cerundolo V (2004) CpG-matured murine plasmacytoid dendritic cells are capable of *in vivo* priming of functional CD8 T cell responses to endogenous but not exogenous antigens. *J Exp Med* 199: 567–579

19 Facchetti F, de Wolf-Peeters C, Mason DY, Pulford K, van den Oord JJ, Desmet VJ (1988) Plasmacytoid T cells. Immunohistochemical evidence for their monocyte/macrophage origin. *Am J Pathol* 133: 15–21

20 Cella M, Jarrossay D, Facchetti F, Alebardi O, Nakajima H, Lanzavecchia A, Colonna M (1999) Plasmacytoid monocytes migrate to inflamed lymph nodes and produce large amounts of type I interferon. *Nat Med* 5: 919–923

21 Kohrgruber N, Groger M, Meraner P, Kriehuber E, Petzelbauer P, Brandt S, Stingl G, Rot A, Maurer D (2004) Plasmacytoid dendritic cell recruitment by immobilized CXCR3 ligands. *J Immunol* 173: 6592–6602

22 Krug A, Uppaluri R, Facchetti F, Dorner BG, Sheehan KC, Schreiber RD, Cella M, Colonna M (2002) IFN-producing cells respond to CXCR3 ligands in the presence of CXCL12 and secrete inflammatory chemokines upon activation. *J Immunol* 169: 6079–6083

23 Kadowaki N, Ho S, Antonenko S, Malefyt RW, Kastelein RA, Bazan F, Liu YJ (2001) Subsets of human dendritic cell precursors express different toll-like receptors and respond to different microbial antigens. *J Exp Med* 194: 863–869

24 Jarrossay D, Napolitani G, Colonna M, Sallusto F, Lanzavecchia A (2001) Specialization and complementarity in microbial molecule recognition by human myeloid and plasmacytoid dendritic cells. *Eur J Immunol* 31: 3388–3393

25 Pulendran B (2004) Immune activation: death, danger and dendritic cells. *Curr Biol* 14: R30–R32

26 Cella M, Scheidegger D, Palmer-Lehmann K, Lane P, Lanzavecchia A, Alber G (1996) Ligation of CD40 on dendritic cells triggers production of high levels of interleukin-12 and enhances T cell stimulatory capacity: T-T help via APC activation. *J Exp Med* 184: 747–752

27 Schulz O, Edwards AD, Schito M, Aliberti J, Manickasingham S, Sher A, Reis E Sousa C (2000) CD40 triggering of heterodimeric IL-12 p70 production by dendritic cells *in vivo* requires a microbial priming signal. *Immunity* 13: 453–462

28 Messi M, Giacchetto I, Nagata K, Lanzavecchia A, Natoli G, Sallusto F (2003) Memory and flexibility of cytokine gene expression as separable properties of human T(H)1 and T(H)2 lymphocytes. *Nat Immunol* 4: 78–86

29 Sallusto F, Palermo B, Lenig D, Miettinen M, Matikainen S, Julkunen I, Forster R, Burgstahler R, Lipp M, Lanzavecchia A (1999) Distinct patterns and kinetics of chemokine production regulate dendritic cell function. *Eur J Immunol* 29: 1617–1625

30 Tang HL, Cyster JG (1999) Chemokine up-regulation and activated T cell attraction by maturing dendritic cells. *Science* 284: 819–822

31 Gunn MD, Tangemann K, Tam C, Cyster JG, Rosen SD, Williams LT (1998) A chemokine expressed in lymphoid high endothelial venules promotes the adhesion and chemotaxis of naive T lymphocytes. *Proc Natl Acad Sci USA* 95: 258–263

32 Verbovetski I, Bychkov H, Trahtemberg U, Shapira I, Hareuveni M, Ben-Tal O, Kutikov I, Gill O, Mevorach D (2002) Opsonization of apoptotic cells by autologous iC3b facilitates clearance by immature dendritic cells, down-regulates DR and CD86, and up-regulates CC chemokine receptor 7. *J Exp Med* 196: 1553–1561

33 Lindquist RL, Shakhar G, Dudziak D, Wardemann H, Eisenreich T, Dustin ML, Nussenzweig MC (2004) Visualizing dendritic cell networks *in vivo*. *Nat Immunol* 5: 1243–1250

34 Hugues S, Fetler L, Bonifaz L, Helft J, Amblard F, Amigorena S (2004) Distinct T cell dynamics in lymph nodes during the induction of tolerance and immunity. *Nat Immunol* 5: 1235–1242

35 Forster R, Schubel A, Breitfeld D, Kremmer E, Renner-Muller I, Wolf E, Lipp M (1999) CCR7 coordinates the primary immune response by establishing functional microenvironments in secondary lymphoid organs. *Cell* 99: 23–33

36 MartIn-Fontecha A, Sebastiani S, Hopken UE, Uguccioni M, Lipp M, Lanzavecchia A, Sallusto F (2003) Regulation of dendritic cell migration to the draining lymph node: impact on T lymphocyte traffic and priming. *J Exp Med* 198: 615–621

37 Robbiani DF, Finch RA, Jager D, Muller WA, Sartorelli AC, Randolph GJ (2000) The leukotriene C(4) transporter MRP1 regulates CCL19 (MIP-3beta, ELC)-dependent mobilization of dendritic cells to lymph nodes. *Cell* 103: 757–768

38 Scandella E, Men Y, Gillessen S, Forster R, Groettrup M (2002) Prostaglandin E2 is a key factor for CCR7 surface expression and migration of monocyte-derived dendritic cells. *Blood* 100: 1354–1361

39 Bakker AB, Hoek RM, Cerwenka A, Blom B, Lucian L, McNeil T, Murray R, Phillips LH, Sedgwick JD, Lanier LL (2000) DAP12-deficient mice fail to develop autoimmunity due to impaired antigen priming. *Immunity* 13: 345–353

40 Bouchon A, Hernandez-Munain C, Cella M, Colonna M (2001) A DAP12-mediated pathway regulates expression of CC chemokine receptor 7 and maturation of human dendritic cells. *J Exp Med* 194: 1111–1122

41. Kissenpfennig A, Henri S, Dubois B, Laplace-Builhe C, Perrin P, Romani N, Tripp CH, Douillard P, Leserman L, Kaiserlian D et al (2005) Dynamics and function of Langerhans cells in vivo dermal dendritic cells colonize lymph node areas distinct from slower migrating Langerhans cells. *Immunity* 22: 643–654

42 Rescigno M, Urbano M, Valzasina B, Francolini M, Rotta G, Bonasio R, Granucci F, Kraehenbuhl JP, Ricciardi-Castagnoli P (2001) Dendritic cells express tight junction proteins and penetrate gut epithelial monolayers to sample bacteria. *Nat Immunol* 2: 361–367

43 Niess JH, Brand S, Gu X, Landsman L, Jung S, McCormick BA, Vyas JM, Boes M,

Ploegh HL, Fox JG et al (2005) CX3CR1-mediated dendritic cell access to the intestinal lumen and bacterial clearance. *Science* 307: 254–258
44 Bonifaz L, Bonnyay D, Mahnke K, Rivera M, Nussenzweig MC, Steinman RM (2002) Efficient targeting of protein antigen to the dendritic cell receptor DEC-205 in the steady state leads to antigen presentation on major histocompatibility complex class I products and peripheral $CD8^+$ T cell tolerance. *J Exp Med* 196: 1627–1638
45 Banchereau J, Palucka AK (2005) Dendritic cells as therapeutic vaccines against cancer. *Nat Rev Immunol* 5: 296–306
46 Martin-Fontecha A, Thomsen LL, Brett S, Gerard C, Lipp M, Lanzavecchia A, Sallusto F (2004) Induced recruitment of NK cells to lymph nodes provides IFN-gamma for T(H)1 priming. *Nat Immunol* 5: 1260–1265
47 Langenkamp A, Messi M, Lanzavecchia A, Sallusto F (2000) Kinetics of dendritic cell activation: impact on priming of TH1, TH2 and nonpolarized T cells. *Nat Immunol* 1: 311–316
48 Sallusto F, Geginat J, Lanzavecchia A (2004) Central memory and effector memory T cell subsets: function, generation, and maintenance. *Annu Rev Immunol* 22: 745–763
49 Adema GJ, de Vries IJ, Punt CJ, Figdor CG (2005) Migration of dendritic cell based cancer vaccines: *in vivo* veritas? *Curr Opin Immunol* 17: 170–174
50 Saiki T, Ezaki T, Ogawa M, Matsuno K (2001) Trafficking of host- and donor-derived dendritic cells in rat cardiac transplantation: allosensitization in the spleen and hepatic nodes. *Transplantation* 71: 1806–1815

Chemokine receptor-mediated signal transduction

Mario Mellado, Carlos Martínez-A. and José Miguel Rodríguez-Frade

Department of Immunology and Oncology, Centro Nacional de Biotecnología/CSIC, UAM Campus de Cantoblanco, E-28049 Madrid, Spain

Introduction

Correct cell movement and positioning are central elements in development, and influence both normal physiology and disease states. Cell movement has probably been studied most extensively in the immune system, where many aspects of the immune response are closely related to coordination of leukocyte trafficking [1–3].

The family of low molecular weight proinflammatory cytokines, termed chemokines, has been implicated directly in governing cell movement. The importance of chemokines in the patterning and plasticity of the immune and nervous systems and in various inflammatory processes has been shown by detection of chemokine/chemokine receptor mRNAs and proteins, use of antagonist molecules, interference RNA or studies of the phenotype of knockout and transgenic animals [4–6].

The chemokines are classified in two main groups. In simplified terms, the inflammatory chemokines recruit cells during inflammatory processes, whereas homeostatic chemokines control haematopoiesis and immune processes in health. In addition to these functional differences, the inflammatory chemokines are inducible and show receptor promiscuity; the homeostatic chemokines are constitutively expressed, with narrow receptor specificity. Examples of the role of chemokines and their receptors in homeostatic processes include regulation of B and T cell homing (CXCR5, CCR7) [7, 8], B cell traffic to mucosa (CCR6) [9] and bone marrow (CXCR4) [10], development of Th1 (CCR5 and CXCR4) and Th2 (CCR3) responses [11], resistance to apoptosis (CXCR5, CCR9) [12], antigen-presenting cell (APC) development (CCR2, CCR8) [13], and dendritic cell (DC) development (CCR6, CCR7, CXCR3) [14].

Since their first description [15], the chemokines have been the subject of great interest due to their potential as targets for drug development in inflammatory diseases. Although recruitment of cell populations and expression of specific chemokines can be correlated in several inflammatory diseases, including asthma

Chemokine Biology – Basic Research and Clinical Application, Volume I
edited by Bernhard Moser, Gordon L. Letts and Kuldeep Neote

[16], bowel disease [17, 18] atherosclerosis [19] or rheumatoid arthritis [20], the redundancy and promiscuity of the chemokine system nonetheless makes it difficult to define the chemokines that are essential in the course of a pathological process.

Classical view of chemokine receptor signalling

Chemokines exert their effects through interactions with seven-transmembrane, G protein-coupled receptors (GPCR) in the target cell membrane [21]. Although similar to many other seven-transmembrane receptors, the chemokine receptors have some unique structural features [22]. Initial studies of chemokine signalling were based in part on information available for other GPCR. Several factors nonetheless slowed chemokine signalling research, including a lack of reliable chemokine-specific reagents and cell-dependent variability in receptor expression. Most studies centred on description of new receptors, assigning ligands to orphan receptors, chemokine-based drug discovery or characterising the chemokine receptors in HIV-1 infection, with limited interest in underlying mechanisms.

The classical view of chemoattractant receptor signalling requires activation of the G protein pathway after chemokine binding [23, 24]. The majority of the responses can be inhibited by pertussis toxin (PTx) treatment, indicating that members of the G_i protein family are the primary transduction partners of these receptors [23, 24]. $G\alpha_i$ associates to the chemokine receptors in response to ligand stimulation; this, and the potent agonist-dependent inhibition of adenylyl cyclase are consistent with receptor coupling to $G\alpha_i$, and mobilisation of intracellular calcium [25, 26]. $G\alpha_i$ is not the only G protein that couples to chemokine receptors; Gq, G16 and G11 also participate in chemokine signalling [27, 28]. Following activation, heterotrimeric G protein dissociates into the $\beta\gamma$ subunit complex and the guanosine triphosphate (GTP)-bound α subunit, each of which is necessary for initiating intracellular signalling responses.

G protein-mediated signalling includes activation of phospholipase C (PLC), resulting in formation of inositol triphosphate [Ins(1,4,5)P3] and diacylglycerol (DAG), responsible for calcium mobilisation and protein kinase C (PKC) activation, respectively [29]. Chemokines also induce activation of phospholipase A_2 (PLA_2) and release of arachidonic acid, which are involved in the chemotactic response, and in triggering of phospholipase D (PLD), which has been implicated in vesicular trafficking and cell transformation in response to chemokines [30, 31].

Through the G protein complex, the chemokine receptor interacts with several signalling pathways. This is the case for the coupling of GPCR kinases (GRK), for which $\beta\gamma$ association with activated GPCR allows formation of a ternary complex with GRK, required for Ser/Thr phosphorylation [26, 32]. The phosphorylated receptor has increased affinity for arrestin-type proteins, whose binding impedes further coupling between the receptor and G proteins, and targets GPCR for internalisation [26, 32].

G protein activity is regulated by altering the transition between GTP- and guanosine diphosphate (GDP)-bound forms, which correspond to active and inactive G protein, respectively. This transition is controlled by regulators of G protein signalling (RGS) proteins that, by acting as GTPase- activating proteins, promote α subunit reassociation with the $\beta\gamma$ complex and prevents its interaction with effectors [33, 34]. Several RGS family members are expressed in lymphocytes, including RGS1, RGS2, RGS10, RGS13, RGS14, RGS16, and RGS19; RGS protein regulates chemotaxis through CXCR2, CXCR4, CXCR5 or CCR3 [35–37].

Chemotaxis requires highly complex motile responses involving changes in cell shape, actin polymerisation/depolymerisation, and cell adhesion [38, 39]. These processes are modulated by guanine nucleotides, and involve regulation by low molecular weight GTP-binding proteins, including Rho, Rac and Cdc42, which modulate actin filament assembly. Chemokine stimulation results in activation of Rho, Rac and Cdc42, which are involved in regulation of focal adhesion, lamellipodia and philopodia, respectively [40–43]. Despite extensive work, the link between these proteins and the chemokine receptors remains unclear.

Phosphatidylinositol-3-kinase (PI3K) activity is rapidly stimulated by chemoattractants. Its role in chemotaxis varies greatly depending on cell type, which may explain the disparity of results reported in the literature [43, 44]. PI3K is activated by GPCR stimulation; this generates 3-phosphorylated lipids that act as second messengers for the downstream effectors PKB, PKC or AKT, as well as for Ras pathways [45–47]. Chemokine-activated PI3K also has a central role in integrin adhesiveness, cell migration and polarisation [43, 44]. Recent data nonetheless implicate DOCK2, a member of the CDM (*Caenorhabditis elegans* CED-5, mammalian Dock180, Mb) regulators of cytoskeleton dynamics protein family, in T and B cell migration [45]. By modulating chemokine-mediated Rac activation, DOCK2 controls T and B cell polarisation and migration in a largely PI3K-independent process; the data thus point to divergent, cell type-dependent functions for DOCK2 and PI3K during chemokine-induced signalling [45]. Chemokines also activate the MAPK (mitogen-activating protein kinase) cascade, which regulates gene expression and modulates cytoskeletal changes necessary for cell migration through pathways involving PLA_2 [48].

Chemokines also activate other tyrosine kinases. Through a molecular complex formed by the focal adhesion kinase (FAK) protein $p125^{FAK}$ and the T cell tyrosine kinase zeta-associated protein (ZAP)-70, CCL5 induces the generation of T cell focal adhesions and subsequent cell activation [49]. Via its SH2 domains, ZAP-70 binds to the phosphotyrosine in the immunoreceptor tyrosine-based activation motif (ITAM) domains of the T cell receptor (TCR) in a process catalysed by $p56^{lck}$ or $p59^{fyn}$ [50]. The link between chemokine signalling with cytoskeletal proteins responsible for migratory and adhesive functions also involves the phosphorylation

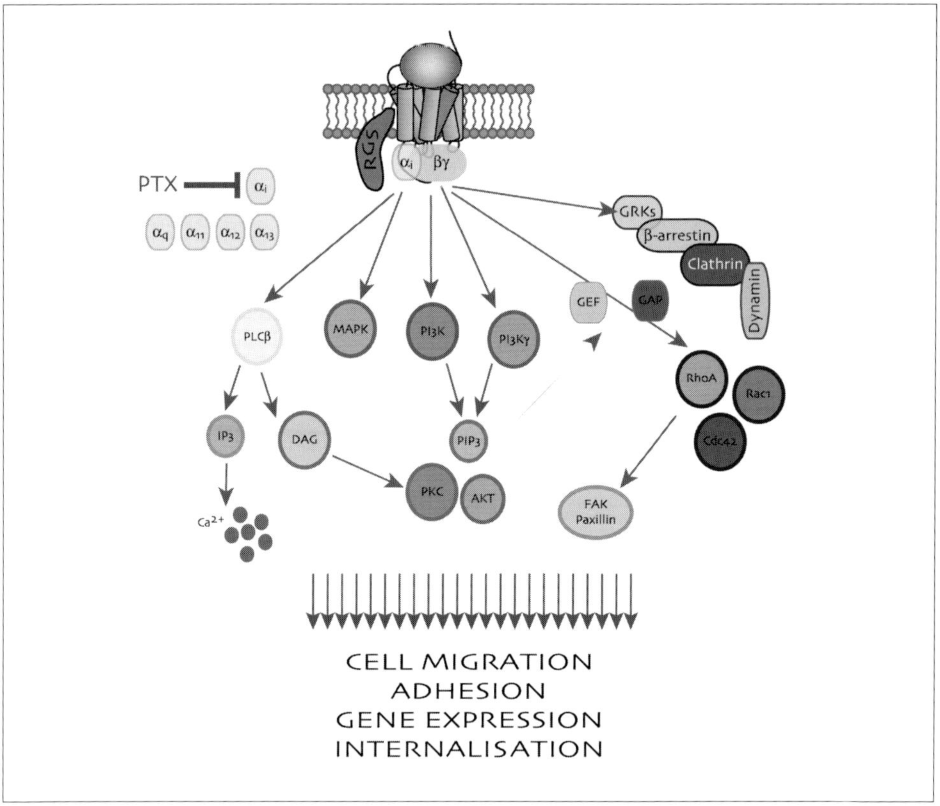

Figure 1
Classical view of chemokine signalling
Following ligand binding, a G protein associates to the receptor; dissociation of its subunits enables activation of several signalling cascades. Abbreviations: PTX, pertussis toxin; RGS, regulator of G-protein signalling; PLC, phospholipase C; IP3, Inositol tri-phosphate; MAPK, mitogen-activated protein kinase; DAG, diacylglycerol; PI3K, phosphatidylinositol 3 kinase; PKC, protein kinase C; PIP3, phosphatidylinositol-3,4,5-triphosphate; GEF, guanine nucleotide exchange factor; GAP, guanine activation protein; FAK, focal adhesion kinase; GRK, G-protein-coupled receptor kinase.

and activation of Pyk2 and subsequent regulation of the JNK/SAPK system [51, 52]. The classical view of the events that follow chemokine/chemokine receptor interaction is summarised in Figure 1.

A realistic view of chemokine signalling must consider the many factors that can modulate chemokine/chemokine receptor expression and function; these include cytokines, co-stimulation, effectors, stress, transformation, pathogens, and mitogens

[53]. Furthermore, although chemokines and their receptors were initially thought to act on specific cell types, we now know that a cell can express varying levels of a number of distinct chemokine receptors, depending on cell cycle status and environmental stimuli. The response of a given cell to a chemokine thus cannot be explained by simple one-receptor/one-chemokine interaction models.

Regulation of chemokine receptor expression and clustering

Events at the cell surface that affect chemokine responses include receptor up- or downregulation, oligomerisation, and their localisation in specialised membrane regions. Receptor regulation is cell-specific; for example, TNF-α/IFN-γ-induced CXCR4 downregulation is reported for neutrophils, but not for monocytes or lymphocytes [54], and H_2O_2 specifically upregulates CCR5 in human monocytes [55]. Another factor is cell status, as is the case of cell cycle-dependent CXCR3 expression [56]. Membrane receptor expression varies greatly in primary cells from one individual to another. This modulation of receptor expression is crucial for a coordinated response to chemokines; some of the many factors that affect it may not always be considered, which explains in part the diversity in results among different laboratories.

The response to a given chemokine depends both on the presence of the appropriate receptor on the target cell, as well as on other mediators that up- or downregulate its expression or the expression of alternative chemokine receptors. Cross-desensitisation has been described, not only for chemokine receptors but also for other GPCR, as is the case of opioid receptors [57]. Chemokine receptors such as CXCR2 can also regulate the functional properties of glutamate receptors [58].

The specialised membrane lipid domains termed rafts also affect individual responses to a given chemokine, as shown by experiments that deplete membrane cholesterol while maintaining other cell functions. Membrane cholesterol is necessary for CXCR4 function, as its depletion inhibits CXCL12 binding and CXCL12-induced Ca^{2+} mobilisation, chemotaxis and cell polarisation [59]. This was also reported for CCR5, whose ligands and even an anti-CCR5 mAb are unable to recognise CCR5 on cholesterol-depleted membranes [60]. Other chemokine functions such as integrin activation also require membrane cholesterol, as shown by the absence of PI3K redistribution in cholesterol-depleted membranes [61].

It has long been known that GPCR can function as oligomers [62]. The current view of GPCR function, which should be also applied to chemokine receptors, is that this family of receptors is found in multiple conformations on the cell surface. Homodimerisation has been demonstrated for CCR2, CCR5 and CXCR4 using co-immunoprecipitation, energy transfer, tagged receptors, and functional assays. Although initially a matter of debate, an increasing number of reports now indicate

Table 1 - Chemokine receptor oligomerisation

Receptor	Ligand	Method	Cell line	Comments	Ref
CCR2/CCR2	+	Co-ipp	T (HEK-293)	First description of chemokine receptor dimerisation.	[63]
		Functional		Relevant for signalling	
		BRET		CCR2 TM peptides diminish BRET signal	[64]
CXCR4/CXCR4	+	BRET		CXCR4 TM peptides diminish BRET signal	[64]
	none	BRET/Co-ipp	T (HEK-293)	No CXCR4/CCR5R5 dimers	[65]
	+	FRET		gp120 increases FRET, but AMD3100 decreases FRET	[66]
		Co-ipp	NT (MOLT4)		[67]
CCR5/CCR5	none	BRET		Anti-CCR5 antibodies increase BRET	[68, 69]
	+		T (HEK-293	Anti-CCR5 mAb also induces dimerisation.	
		Co-ipp		Block HIV-1 infection	[70]
	ND		T (HeLa-P4/ HEK-293)	Trans-dominant effect on cell membrane expression	[71]
	+	NFRET	T (HEK-293/ I1.2)	Ligands stabilize preformed receptor dimers.	[72]
		FLIM		TM1-TM4 peptides diminish FRET signals and block function	
		Co-ipp	NT (human PBL)		
		Functional			
		Bio-informatic			
CXCR2/CXCR2	none	Co-ipp	T (HEK-293)	GluR1 co-expression impairs dimer formation	[73]
		Function	NT (cerebellar granule neurons)	CXCR2 deletion mutants act as dominant negative receptors	
CCR2/CCR5	–	Co-ipp	T (CHO-K1)	Cross-competition in ligand binding assays	[74]
		BRET			
	+	FRET-FLIM	T (HEK-293)	Coupling of distinct signalling molecules to homo- or heterodimers	[75]
		Functional	NT (human PBMC)		

Table 1 - continued

Receptor	Ligand	Method	Cell line	Comments	Ref
CCR2/CCR5	ND	*In silico* (lipid-facing mutational analysis)	ND	Bio-informatic analysis indicating the feasibility of CCR2/CCR5 heterodimers	[76]
	+	Co-ipp FRET	T (HEK-293) NT (Mono Mac 1)	Ligands and mAb stabilise preformed receptor homo- and heterodimers. Blocks HIV-1 entry	[77, 78]
CCR5/ CXCR4	+	Co-ipp	T (HEK-293)	Ligands induce receptor dimers. Blocks HIV-1 entry	[78]
	ND	Co-ipp colocalisation	T (NIH 3T3)	CCR5 interference on CXCR4 expression, endocytosis and HIV-1 co-receptor activity.	[79]
CCR2/ CXCR4	ND	Co-ipp FRET	T (HEK-293) NT (Mono Mac 1)	Anti-CCR2 mAb stabilise preformed receptor homo- and heterodimers. Blocks HIV-1 entry	[77]
	+/–	BRET	T (HEK-293)	CXCR4 TM peptides do not affect BRET signal	[64]
CCR5/μ, κ, δ-OR	+	Co-ipp crosslinking	NT (CEMx174)	Morphine alters chemokine function	[80]
CCR5/m-OR	none	Co-ipp	T (CHO)	Cross-desensitisation at receptor and post-receptor level	[81]
CXCR2/GluR1	ND	Co-ipp colocalisation	T (HEK-293) NT (granule neurons)	GluR1 co-expression inhibits CXCL2 function	[58]

Abbreviations: +, stabilises; –, diminishes; Co-ipp, co-immunoprecipitation; BRET, bioluminescence resonance energy transfer; FRET, fluorescence resonance energy transfer; N-FRET, normalized FRET; FLIM, fluorescence life-time imaging microscopy; T, transfected; NT, non-transfected; TM, transmembrane; ND, not done the relevance of oligomerisation for chemokine function. Interaction between different chemokine receptors has also been reported, as has interaction between chemokine receptors and other GPCR, such as opioid receptors. A summary of chemokine receptors known to dimerise, as well as the functional consequences, is shown in Table 1.

As it becomes clearer that multiple chemokine receptor conformations are found on the cell membrane, controversy has moved to the definition of ligand function in promoting or altering these oligomers, also with regard to functional consequences. Change in the equilibrium between monomers and oligomers is assumed to be part of the activation process for many receptors. Three situations emerge from studies carried out to date: (1) dimers are detected and are not affected by ligand stimulation, (2) dimers are detected and ligand stimulation modulates their presence, or (3) dimers are not detected in the absence of ligand. Although further experiments are required to address these issues properly, initial data indicate that chemokines stabilise a preformed receptor conformation to initiate the signalling cascade [62, 82, 83].

These findings not only confirm chemokine receptor homo- and heterodimerisation, but also suggest that GPCR oligomer assemblies have a number of functional consequences. In analogy, γ-aminobutyric acid (GABA) or vasopressin receptor dimerisation favours receptor entry in the export system, thus influencing receptor trafficking; this is consistent with GPCR dimerisation in the endoplasmic reticulum (ER) [84]. The lack of cell surface CCR5 expression in CCR5Δ32 heterozygous individuals is suggested to be due to ER retention of CCR5-CCR5Δ32 heterodimers [71]. The role of ligand in promoting or inhibiting receptor oligomerisation is a central question, and a consensus has not been reached. Several studies suggest that ligand stabilises or promotes receptor dimers, whereas others indicate the pre-existence of oligomer; these differences may reflect difficulties in interpreting results derived from distinct analytical techniques (see Tab. 1). For example, Western blot analyses indicate ligand-induced CCR2 dimerisation, whereas data from energy transfer techniques limit the role of ligand to stabilisation of pre-existing homo- and heterodimers [63, 72].

Oligomerisation would also explain some of the pharmacological properties of GPCR, as well as some reported differences in signal transduction and receptor internalisation [74, 75, 85]. As there are still relatively few studies of chemokine receptor dimerisation, it is nonetheless difficult to form a clear view of the functional consequences. Different laboratories describe changes in G protein coupling, synergistic effects, or negative cooperation between chemokine receptors, as well as between chemokine and opioid receptors; it is nonetheless clear that at least some chemokine receptors form homo- and heterodimers, and that functional read-out varies as a consequence of activating distinct receptor conformations [62]. Blocking dimerisation has been shown to impede receptor function both *in vivo* and *in vitro* [72].

Chemokine receptor interactions affect chemokine-mediated signal transduction

Signalling through chemokine receptors has been assumed to be almost exclusively G-protein mediated, although chemokines also promote an increase in tyrosine kinase (TK) activity [86]. CXC chemokine activation of the src-related lyn TK was reported in human neutrophils, and CXCL1, CXCL7 or CXCL8 binding in human neutrophils triggers a rapid, time-dependent increase in the tyrosine autophosphorylating activity of the lyn kinase [87]. As discussed above, various kinases participate in late chemokine signalling events. Hints of a role for TK pathways in early signalling were provided by a report on PTx-independent tyrosine phosphorylation of CCR2 [86, 88]; this early phosphorylation is induced by Janus (JAK) kinases, whose activation is nearly simultaneous with their association to the chemokine receptor. Similar results were later reported for CCR5, CCR7 and CXCR4 receptors [67, 69, 89]. Although JAK involvement in chemokine signalling was unexpected, chemokine receptors are not the only GPCR known to activate JAK, as exemplified by the angiotensin type 1 and thyroid hormone receptors [90, 91]. In contrast to their binding to cytokine receptors, JAK are not constitutively associated to GPCR; this can be explained by the lack of JAK-binding consensus sequences in GPCR [91]. In addition, JAK association and activation seem quite variable in GPCR, and there are as yet no common rules for predicting the nature of this interaction [90, 91].

As for most new findings, chemokine-mediated JAK activation is debated. CXCL12 induces neither migration nor calcium mobilisation in JAK-deficient cells, or in the same cells reconstituted with a kinase-dead mutant of JAK, and no G_i association to the receptor was found in these cells [92]. These data are consistent with the effect of JAK inhibitors on chemokine function and G_i coupling to CCR2 [67, 93]. The CXCR4 cytoplasmic domains involved in JAK2 and signal transducers and transactivators of transcription (STAT)3 phosphorylation were recently described, and involve residues in the third intracellular loop [94], although more detailed analyses are needed to determine whether this is a conserved feature in all chemokine receptors. JAK activation through chemokine receptors has also been shown *in vivo*, as JAK blockade affects CCR7-mediated cell rolling [89]. Chemokine-mediated JAK activation is also fundamental for crosstalk with other key mediators of leukocyte function, such as cytokines and growth factors, via mechanisms that involve members of the suppressor of cytokine signalling (SOCS) family [92, 93, 95]. SOCS proteins are upregulated through cytokine-induced, JAK/STAT-mediated pathways and regulate cytokine signalling by binding to the receptor or to JAK, blocking JAK activation [95]. SOCS are also induced by other proteins that activate STAT, such as chemokines [92, 93]. Cytokine/growth factor- or chemokine-upregulated SOCS are available to bind to both receptor types, allowing intracellular communication between these receptor families [92, 93]. This novel view of chemokine signalling is summarised in Figure 2.

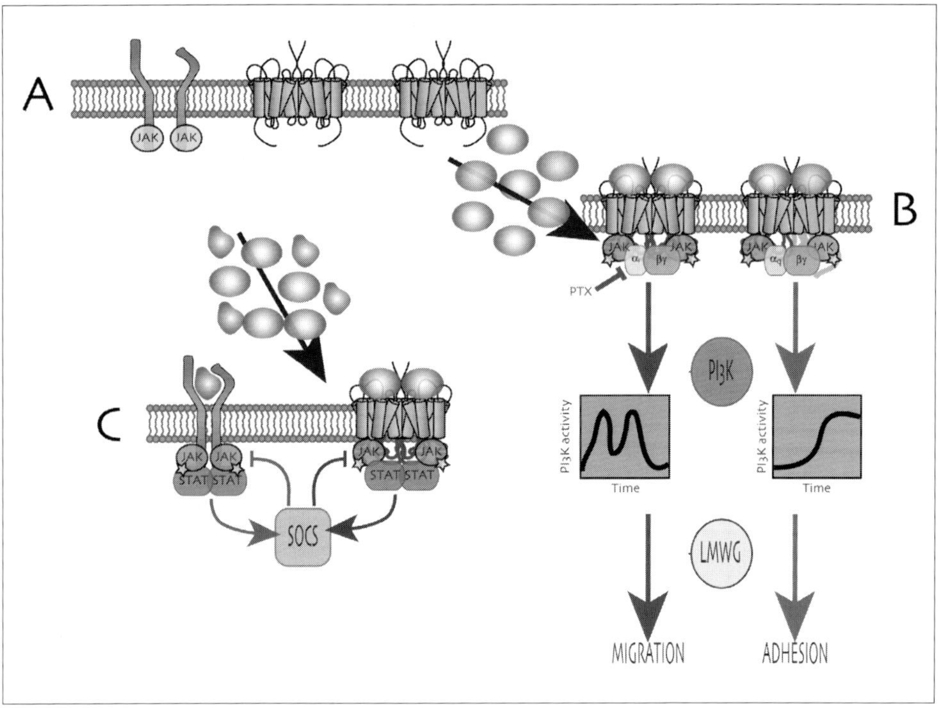

Figure 2
Alternative view of chemokine signalling
Chemokines are present on the cell surface in multiple conformations, together with other cell membrane proteins such as growth factor and cytokine receptors (A). Ligands stabilise active chemokine receptor conformations that include the same (homodimers) or different receptors (heterodimers), resulting in the differential activation of signalling pathways (B). When cytokine/growth factor and chemokine receptors are activated through JAK/STAT activation, members of the SOCS protein family are upregulated, resulting in a checkpoint for cross-regulation of both families of cellular mediators. Abbreviations: JAK, Janus kinase; STAT, signal transducer and activator of transcription; SOCS, suppressors of cytokine signalling; PTX, pertussis toxin; PI3K, phosphatidylinositol 3 kinase; LMWG, low molecular weight GTP binding proteins.

In this view, chemokine signalling would be initiated by ligand-mediated stabilisation of a multimeric receptor conformation, which would allow JAK association and activation, followed by G protein coupling to the receptor. These three steps are critical for chemokine function, and blockade of dimer formation, JAK activity or G protein coupling severely impairs chemokine function. Many combinations of signalling pathways nonetheless remain to be explored.

Conclusions

Despite their newly-found importance in numerous pathophysiological situations, the chemokines behave like many other well-known GPCR ligands. Concepts that are still not settled in the chemokine field were resolved long ago for other GPCR. The therapeutically promising vision of one-chemokine/one-receptor/one-cell type has been replaced by a much more complex view that includes chemokine promiscuity with distinct receptors in various possible conformations. In addition, these molecules are expressed in various cell types, depending on their differentiation or activation status. The outcome is a vast array of possible cell responses as the result of receptor routing into distinct signalling pathways.

Current methods for interference with chemokine function include modification of receptor expression, chemokine sequestration or chemokine blockade. Modification of chemokine-activated signalling pathways presents an attractive target for therapeutic intervention, although a number of questions remain to be answered. If a chemokine receptor can exist in several conformations, the contribution of each conformation to receptor function must be evaluated, including that of monomers. The number of receptors that must be engaged to trigger a given cell response must be established, to determine how many are to be targeted for effective blockade of chemokine function.

Specific TM1 and TM4 residues have recently been implicated in CCR5 receptor dimerisation [72]. Studies are needed for each receptor to ascertain the specific regions involved in oligomer stabilisation. Synthetic peptides that impede chemokine function by blocking receptor dimerisation could be used to develop molecules that interfere with receptor function. Another important issue is the identification of appropriate target receptors; since different receptor types can interact, it must be assured that the desired signalling event is stimulated or repressed. Oligomerisation has been reported for a representative, but still reduced number of chemokine receptors (see Tab. 1). We await a full inventory of the receptors that homodimerise and those that heterodimerise, as well as a list of all possible partners in the latter case. Bispecific reagents that stabilise specific receptor conformations may also represent a line of future research. Finally, therapeutic translation of these concepts must also consider the relationships between chemokine receptors and non-chemotactic GPCR, and with cytokine and growth factor receptors that share common signalling pathways; these interactions, whether natural or provoked, could lead to chemokine receptor non-responsiveness.

Reports describing these interactions are constantly increasing, and in the near future we hope to have a clearer view of the functions of chemokine/receptor groups in physiological and pathological processes.

Acknowledgments

We thank the members of the DIO chemokine group, who contributed to some of the work described. We also thank C. Bastos and C. Mark for secretarial and editorial assistance, respectively. The Department of Immunology and Oncology was founded and is supported by the Spanish Council for Scientific Research (CSIC) and by Pfizer.

References

1 Cyster JG (2003) Lymphoid organ development and cell migration. *Immunol Rev* 195: 5–14

2 Petrie HT (2003) Cell migration and the control of post-natal T-cell lymphopoiesis in the thymus. *Nat Rev Immunol* 3: 859–866

3 Moser B, Wolf M, Loetscher P (2004) Chemokine: Multiple levels of leukocyte migration control. *Trends Immunol* 25: 75–84

4 Carter PH (2002) Chemokine receptor antagonism as an approach to anti-inflammatory theraphy: 'just right' or plain wrong? *Curr Opin Chem Biol* 6: 510–525

5 Power CA (2003) Knock out models to dissect chemokine receptor function *in vivo*. *J Immunol Methods* 273: 73–82

6 Houshmand P, Zlotnik A (2003) Therapeutic applications in the chemokine superfamily. *Curr Opin Chem Biol* 7: 457–460

7 Schaerli P, Willimann K, Lang AB, Lipp M, Loetscher P, Moser B (2000) CXC chemokine receptor 5 expression defines follicular homing T cells with B cell helper function. *J Exp Med* 192: 1553–1562

8 Henning G, Ohl L, Junt T, Reiterer P, Brinkmann V, Nakano H, Hohenberger W, Lipp M, Forster R (2001) CC chemokine receptor 7-dependent and -independent pathways for lymphocyte homing: modulation by FTY720. *J Exp Med* 194: 1875–1881

9 Williams IR (2004) Chemokine receptors and leukocyte trafficking in the mucosal immune system. *Immunol Res* 29: 283–292

10 Ma Q, Jones D, Borghesani PR, Segal RA, Nagasawa T, Kishimoto T, Bronson RT, Springer TA (1998) Impaired B-lymphopoiesis, myelopoiesis, and derailed cerebellar neuron migration in CXCR4- and SDF-1-deficient mice. *Proc Natl Acad Sci USA* 95: 9448–9453

11 Sallusto F, Lanzavecchia A (2000) Understanding dendritic cell and T-lymphocyte traffic through the analysis of chemokine receptor expression. *Immunol Rev* 177: 134–140

12 Youn BS, Yu KY, Oh J, Lee J, Lee TH, Broxmeyer HE (2002) Role of the CC chemokine receptor 9/TECK interaction in apoptosis. *Apoptosis* 7: 271–276

13 Dieu-Nosjean MC, Vicari A, Lebecque S, Caux C (1999). Regulation of dendritic cell trafficking: a process that involves the participation of selective chemokines. *J Leukoc Biol* 66: 252–262

14 Martin-Fontecha A, Sebastiani S, Hopken UE, Uguccioni M, Lipp M, Lanzavecchia A, Sallusto F (2003) Regulation of dendritic cell migration to the draining lymph node: impact on T lymphocyte traffic and priming. *J Exp Med* 198: 615–662

15 Oppenheim JJ, Howard OMZ, Goetzl E (2000) Chemotactic factors, neuropeptides, and other ligands for seven transmembrane receptors. In: Oppenheim JJ, Feldmann M (eds): *Cytokine Reference*. Academic Press, London, 985–1021
16 Lukacs NW, Miller AL, Hogaboam CM (2003) Chemokine receptors in asthma: searching for the correct immune targets. *J Immunol* 170: 11–15
17 Papadakis KA, Landers C, Prehn J, Kouroumalis EA, Moreno ST, Gutierrez-Ramos JC, Hodge MR, Targan SR (2003) CC chemokine receptor 9 expression defines a subset of peripheral blood lymphocytes with mucosal T cell phenotype and Th1 or T-regulatory 1 cytokine profile. *J Immunol* 171: 159–165
18 Scheerens H, Hessel E, de Waal-Malefyt R, Leach MW, Rennick D (2001) Characterization of chemokines and chemokine receptors in two murine models of inflammatory bowel disease: IL-$10^{-/-}$ mice and Rag-$2^{-/-}$ mice reconstituted with $CD4^{+}CD45RB^{high}$ T cells. *Eur J Immunol* 31: 1465–1474
19 Lesnik P, Haskell CA, Charo IF (2003) Decreased atherosclerosis in $CX3CR1^{-/-}$ mice reveals a role for fractalkine in atherogenesis. *J Clin Invest* 111: 333–340
20 Garcia-Vicuna R, Gomez-Gaviro MV, Dominguez-Luis MJ, Pec MK, Gonzalez-Alvaro I, Alvaro-Gracia JM, Diaz-Gonzalez F (2004) CC and CXC chemokine receptors mediate migration, proliferation, and matrix metalloproteinase production by fibroblast-like synoviocytes from rheumatoid arthritis patients. *Arthritis Rheum* 50: 3866–3877
21 Horuk R (1994) The interleukin-8-receptor family: from chemokines to malaria. *Immunol Today* 15: 169–174
22 Murphy PM (1994) The molecular biology of leukocyte chemoattractant receptors. *Annu Rev Immunol* 12: 593–633
23 Wu D, LaRosa GJ, Simon MI (1993) G protein-coupled signal transduction pathways for interleukin-8. *Science* 261: 101–103
24 Kuang Y, Wu Y, Jiang H, Wu D (1996) Selective G protein coupling by C-C chemokine receptors. *J Biol Chem* 271: 3975–3978
25 Damaj BB, McColl SR, Mahana W, Crouch MF, Naccache PH (1996) Physical association of Gi2alpha with interleukin-8 receptors. *J Biol Chem* 271: 12783–12789
26 Aragay AM, Mellado M, R-Frade JM, Martin AM, Jimenez-Sainz MC, Martínez-A C, Mayor Jr F (1998) Monocyte chemoattractant protein-1-induced CCR2B receptor desensitization mediated by the G protein-coupled receptor kinase 2. *Proc Natl Acad Sci USA* 95: 2985–2990
27 Arai H, Charo IF (1996) Differential regulation of G-protein-mediated signaling by chemokine receptors. *J Biol Chem* 271: 21814–21819
28 Mellado M, Rodriguez-Frade JM, Vila-Coro AJ, Fernandez S, Martin de Ana A, Jones DR, Toran JL, Martinez-A C (2001) Chemokine receptor homo- or heterodimerization activates distinct signaling pathways. *EMBO J* 20: 2497–2507
29 Li Z, Jiang H, Xie W, Zhang Z, Smrcka AV, Wu D (2000) Roles of PLC-β2 and-β3 and PI3Kγ in chemoattractant-mediated signal transduction. *Science* 287: 1046–1049
30 Carnevale KA, Cathcart MK (2001) Calcium-independent phospholipase A2 is required

for human monocyte chemotaxis to monocyte chemoattractant protein 1. *J Immunol* 167: 3414–3421

31 Bacon KB, Schall TJ, Dairaghi DJ (1998) RANTES activation of phospholipase D in Jurkat T cells: requirement of GTP-binding proteins ARF and RhoA. *J Immunol* 160: 1894–1900

32 Franci C, Gosling J, Tsou CL, Coughlin SR, Charo IF (1996) Phosphorylation by a G protein-coupled kinase inhibits signaling and promotes internalization of the monocyte chemoattractant protein-1 receptor. Critical role of carboxyl-tail serines/threonines in receptor function *J Immunol* 157: 5606–5612

33 Kehrl JH (1998) Heterotrimeric G protein signaling: roles in immune function and fine-tuning by RGS proteins. *Immunity* 8: 1–10

34 Moratz C, Harrison K, Kehrl JH (2004) Regulation of chemokine-induced lymphocyte migration by RGS proteins. *Methods Enzymol* 389: 15–32

35 Bowman EP, Campbell JJ, Druey KM, Scheschonka A, Kehrl JH, Butcher EC (1998) Regulation of chemotactic and proadhesive responses to chemoattractant receptors by RGS (regulator of G-protein signaling) family members. *J Biol Chem* 273: 28040–28048

36 Shi GX, Harrison K, Wilson GL, Moratz C, Kehrl JH (2002) RGS13 regulates germinal center B lymphocytes responsiveness to CXC chemokine ligand (CXCL)12 and CXCL13. *J Immunol* 169: 2507–2515

37 Lippert E, Yowe DL, Gonzalo JA, Justice JP, Webster JM, Fedyk ER, Hodge M, Miller C, Gutierrez-Ramos JC, Borrego F et al (2003) Role of regulator of G protein signaling 16 in inflammation-induced T lymphocyte migration and activation. *J Immunol* 171: 1542–1555

38 del Pozo MA, Nieto M, Serrador JM, Sancho D, Vicente-Manzanares M, Martinez C, Sanchez-Madrid F (1998) The two poles of the lymphocyte: specialized cell compartments for migration and recruitment. *Cell Adhes Común* 6: 125–133

39 Laudanna C, Kim JY, Constantin G, Butcher E (2002) Rapid leukocyte integrin activation by chemokines. *Immunol Rev* 186: 37–46

40 Laudanna C, Campbell JJ, Butcher EC (1996) Role of Rho in chemoattractant-activated leukocyte adhesion through integrins. *Science* 271: 981–983

41 Takesono A, Horai R, Mandai M, Dombroski D, Schwartzberg PL (2004) Requirement for Tec kinases in chemokine-induced migration and activation of Cdc42 and Rac. *Curr Biol* 14: 917–922

42 Haddad E, Zugaza JL, Louache F, Debili N, Crouin C, Schwarz K, Fischer A, Vainchenker W, Bertoglio J (2001) The interaction between Cdc42 and WASP is required for SDF-1-induced T-lymphocyte chemotaxis. *Blood* 97: 33–38

43 Weiss-Haljiti C, Pasquali C, Ji H, Gillieron C, Chabert C, Curchod ML, Hirsch E, Ridley AJ, van Huijsduijnen RH, Camps M, Rommel C (2004) Involvement of phosphoinositide 3-kinase gamma, Rac, and PAK signaling in chemokine-induced macrophage migration. *J Biol Chem* 279: 43273–43284

44 Sasaki T, Irie-Sasaki J, Jones R, Oliveira-dos Santos A, Standford W, Bolon B, Wakeman

A, Itie A, Bouchard D, Kzieradki I et al (2000) Function of PI3Kγ in thymocyte development, T cell activation, and neutrophil migration. *Science* 287: 1040–1046
45 Nombela-Arrieta C, Lacalle RA, Montoya MC, Kunisaki Y, Megias D, Marques M, Carrera AC, Manes S, Fukui Y, Martinez-A C, Stein JV (2004) Differential Requirements for DOCK2 and phosphoinositide-3-kinase gamma during T and B lymphocyte homing. *Immunity* 21: 429–441
46 Bonacchi A, Romagnani P, Romanelli RG, Efsen E, Annunziato F, Lasagni L, Francalanci M, Serio M, Laffi G, Piunzani M et al (2001) Signal transduction by the chemokine receptor CXCR3 Activation of Ras/ERK, Src and phosphatidylinositol 3-kinase/Akt controls cell migration and proliferation in human vascular pericytes. *J Biol Chem* 276: 9945–9954
47 Jimenez C, Armas Portela R, Mellado M, Rodriguez-Frade JM, Collard J, Serrano A, Martinez-A C, Avila J, Carrera AC (2000) Role of the PI3K regulatory subunit in the control of actin organization and cell migration. *J Cell Biol* 151: 249–261
48 Smit MJ, Verdijk P, van der Raaij-Helmer EM, Navis M, Hensbergen PJ, Laurs R, Tensen CP (2003) CXCR3-mediated chemotaxis of human T cells is regulated by a Gi- and phospholipase C-dependent pathway and not via activation of MEK/p44/p42 MAPK nor Akt/PI-3 kinase. *Blood* 102: 1959–1965
49 Bacon KB, Szabo MC, Yssel H, Bolen JB, Schall TJ (1996) RANTES induces tyrosine kinase activity of stably complexed $p125^{FAK}$ and ZAP-70 in human T cells. *J Exp Med* 184: 873–882
50 Chan AC, Desai DM, Weiss A (1994) The role of protein tyrosine kinases and protein tyrosine phosphatases in T cell antigen receptor signal transduction. *Annu Rev Immunol* 12: 555–592
51 Ganju RK, Dutt P, Wu L, Newman W, Avraham H, Avraham S, Groopman JE (1998) β-chemokine receptor CCR5 signals via the novel tyrosine kinase RAFTK. *Blood* 91: 791–797
52 Okigaki M, Davis C, Falasca M, Harroch S, Felsenfeld DP, Sheetz MP, Schlessinger J (2003) Pyk2 regulates multiple signaling events crucial for macrophage morphology and migration. *Proc Natl Acad Sci USA* 100: 10740–10745
53 Locati M, Otero K, Schioppa T, Signorelli P, Perrier P, Baviera S, Sozzani S, Mantovani A (2002) The chemokine system: tuning and shaping by regulation of receptor expression and coupling in polarized responses. *Allergy* 57: 972–982
54 Brühl H, Cohen CD, Linder S, Kretzler M, Schlöndorff D, Mack M (2003) Post-translational and cell type-specific regulation of CXCR4 expression by cytokines. *Eur J Immunol* 33: 3028–3037
55 Saccani A, Saccani S, Orlando S, Sironi M, Bernasconi S, Ghezzi P, Mantovani A, Sica A (2000) Redox regulation of chemokine receptor expression. *Proc Natl Acad Sci USA* 97: 2761–2766
56 Romagnani P, Annunziato F, Lasagni L, Lazzeri E, Beltrame C, Francalanci M, Uguccioni M, Galli G, Cosmi L, Maurenzig L et al (2001) Cell cycle-dependent expression of

CXC chemokine receptor 3 by endothelial cells mediates angiostatic activity. *J Clin Invest* 107: 53–63

57 Szabo I, Chen XH, Xin L, Adler MW, Howard OM, Oppenheim JJ, Rogers TJ (2002) Heterologous desensitization of opioid receptors by chemokines inhibits chemotaxis and enhances the perception of pain. *Proc Natl Acad Sci USA* 99: 10276–10281

58 Limatola C, Di Bartolomeo S, Trettel F, Lauro C, Ciotti MT, Mercanti D, Castellani L, Eusebi F (2003) Expression of AMPA-type glutamate receptors in HEK cells and cerebellar granule neurons impairs CXCL2-mediated chemotaxis. *J Neuroimmunol* 134: 61–71

59 Nguyen DH, Taub D (2002) CXCR4 function requires membrane cholesterol: implications for HIV infection. *J Immunol* 168: 4121–4126

60 Nguyen DH, Taub D (2002) Cholesterol is essential for macrophage inflammatory protein 1β binding and conformational integrity of CC chemokine. *Blood* 99: 4298–4306

61 Mañes S, Lacalle RA, Gómez-Moutón C, Martínez-A C (2003) From rafts to crafts: membrane asymmetry in moving cells. *Trends Immunol* 24: 319–325

62 Terrillon S, Bouvier M (2004) Roles of G-protein-coupled receptor dimerization. *EMBO Rep* 5: 30–34

63 Rodríguez-Frade JM, Vila-Coro A, Martin de Ana A, Albar JP, Martínez-A C, Mellado M (1999) The chemokine monocyte chemoattractant protein-1 induces functional responses through dimerization of its receptor. *Proc Natl Acad Sci USA* 96: 3628–3633

64 Percherancier Y, Berchiche Y, Slight I, Volkmer-Engert R, Tamamura H, Fujii N, Bouvier M, Heveker N (2005) Bioluminescence resonance energy transfer reveals ligand-induced conformational changes in CXCR4 homo- and heterodimers. *J Biol Chem* PMID 15632118

65 Babcock GJ, Farzan M, Sodroski J (2003) Ligand-independent dimerization of CXCR4, a principal HIV-1 coreceptor. *J Biol Chem* 278: 3378–3385

66 Toth PT, Ren D, Miller RJ (2004) Regulation of CXCR4 dimerization by the chemokine SDF-1α and the HIV-1 coat protein gp120: a fluorescence resonance energy transfer (FRET) study. *J Pharmacol Exp Ther* 310: 8–17

67 Vila-Coro AJ, Rodríguez-Frade JM, Martín de Ana A, Moreno-Ortíz MC, Martínez-A C, Mellado M (1999) The chemokine SDF-1α triggers CXCR4 receptor dimerization and activates the JAK/STAT pathway. *FASEB J* 13: 1699–1710

68 Issafras H, Angers S, Bulenger S, Blanpain C, Parmentier M, Labbe-Jullie C, Bouvier M, Marullo S (2002) Constitutive agonist-independent CCR5 oligomerization and antibody-mediated clustering occurring at physiological levels of receptors. *J Biol Chem* 277: 34666–34673

69 Rodriguez-Frade JM, Vila-Coro A, Martin de Ana A, Nieto M, Sánchez-Madrid F, Proudfoot AEI, Wells TNC, Martínez-A C, Mellado M (1999) Similarities and differences in RANTES- and (AOP)-RANTES-triggered signals: implications for chemotaxis M. *J Cell Biol* 144: 755–765

70 Mellado M, Vila-Coro AJ, Martín de Ana A, Lucas P, Del Real G, Martínez-A C,

Rodríguez-Frade JM (2000) HIV-1 infection through the CCR5 receptor is blocked by receptor dimerization. *Proc Natl Acad Sci USA* 97: 3388–3393
71 Chelli M, Alizon M (2001) Determinants of the trans-dominant negative effect of truncated forms of the CCR5 chemokine receptor. *J Biol Chem* 276: 46975–46982
72 Hernanz-Falcón P, Rodríguez-Frade JM, Serrano A, Juan D, del Sol A, Soriano SF, Roncal F, Gómez L, Valencia A, Martínez-A C et al (2004) Identification of amino acid residues critical for chemokine receptor dimerization. *Nat Immunol* 5: 216–223
73 Trettel F, Di Bartolomeo S, Lauro C, Catalano M, Ciotti MT, Limatola C (2003) Ligand-independent CXCR2 dimerization. *J Biol Chem* 278: 40980–40988
74 El-Asmar L, Springael JY, Ballet S, Andrieu EU, Vassart G, Parmentier M (2004) Evidence for negative cooperativity within CCR5-CCR2b heterodimers. *Mol Pharmacol* 67: 460–469
75 Mellado M, Rodríguez-Frade JM, Vila-Coro AJ, Martín de Ana AM, Jones DR, Martínez-A C (2001) Chemokine receptor homo- or heterodimerization activates distinct signaling pathways: implications for increased sensitivity and dynamic range of chemotaxis. *EMBO J* 20: 2497–2507
76 Gouldson PR, Dean MK, Snell CR, Bywater RP, Gkoutos G, Reynolds CA (2001) Lipid-facing correlated mutations and dimerization in G protein coupled receptors. *Protein Eng* 14: 759–767
77 Rodríguez-Frade JM, del Real G, Serrano A, Hernanz-Falcón P, Soriano SF, Vila-Coro AJ, Martín de Ana A, Lucas P, Prieto I, Martínez-A C et al (2004) Blocking HIV-1 infection via CCR5 and CXCR4 receptors by acting in trans on the CCR2 chemokine receptor. *EMBO J* 23: 66–76
78 Mellado M, Rodríguez-Frade JM, Vila-Coro AJ, de Ana AM, Martínez-A C (1999) Chemokine control of HIV-1 infection. *Nature* 400: 723–724
79 Wang J, Alvarez R, Roderiquez G, Guan E, Norcross MA (2004) Constitutive association of cell surface CCR5 and CXCR4 in the presence of CD4. *J Cell Biochem* 93: 753–760
80 Suzuki S, Chuang JF, Yau P, Doi RH, Chuang RY (2002) Interactions of opioid and chemokine receptors: oligomerization of mu, kappa and delta with CCR5 on immune cells. *Exp Cell Res* 280: 192–200
81 Chen C, Li J, Bot G, Szabo I, Rogers TJ, Liu-Chen LY (2004) Heterodimerization and cross-desensitization between the mu-opioid receptor and the chemokine CCR5 receptor. *Eur J Pharmacol* 483: 175–186
82 George SR, O'Dowd BF, Lee SP (2002) G-protein-coupled receptor oligomerization and its potential for drug discovery. *Nat Rev Drug Discov* 1: 808–820
83 Bulenger S, Marullo S, Bouvier M (2005) Emerging role of homo- and heterodimerization in G-protein-coupled receptor biosynthesis and maturation. *Trends Pharmacol Sci* 26: 131–137
84 Bockaert J, Pin JP (1999) Molecular tinkering of G protein-coupled receptors: an evolutionary success. *EMBO J* 18: 1723–1729
85 Gomes I, Jordan BA, Gupta A, Rios C, Trapaidze N, Devi LA (2001) G protein coupled

receptor dimerization: implications in modulating receptor function. *J Mol Med* 79: 226–242

86 Mellado M, Rodriguez-Frade JM, Aragay A, del Real G, Martin AM, Vila-Coro AJ, Serrano A, Mayor F Jr, Martinez-A C (1998) The chemokine monocyte chemotactic protein 1 triggers Janus Kinase 2 activation and tyrosine phosphorylation of the CCR2B receptor. *J Immunol* 161: 805–813

87 Gaudry M, Gilbert C, Barabé F, Poubelle PE, Naccache PH (1995) Activation of Lyn is a common element of the stimulation of human neutrophils by soluble and particulate agonists. *Blood* 86: 3567–3574

88 Mueller A, Stange PG (2004) CCL3, acting via the chemokine receptor CCR5, leads to independent activation of Janus kinase 2 (JAK2) and Gi proteins. *FEBS Lett* 570: 126–132

89 Stein JV, Soriano SF, M'rini C, Nombela-Arrieta C, de Buitrago GG, Rodriguez-Frade JM, Mellado M, Girard JP, Martinez-A C (2003) CCR7-mediated physiological lymphocyte homing involves activation of a tyrosine kinase pathway. *Blood* 101: 38–44

90 Park ES, Kim H, Suh JM, Park SJ, You SH, Chung HK, Lee KW, Kwon OY, Cho BY, Kim YK et al (2000) Involvement of JAK/STAT (Janus kinase/signal transducer and activator of transcription) in the thyrotropin signaling pathway. *Mol Endocrinol* 14: 662–670

91 Ali MS, Sayeski PP, Dirksen LB, Hayzer DJ, Marrero MB, Bernstein KE (1997) Dependence on the motif YIPP for the physical association of JAK2 kinase with the intracellular carboxy tail of the angiotensin II AT1 receptor. *J Biol Chem* 272: 23382–23388

92 Soriano SF, Serrano A, Hernanz-Falcon P, Martín de Ana A, Monterrubio M, Martinez-A C, Rodríguez-Frade JM, Mellado M (2003) Chemokines integrate JAK/STAT and G-protein pathways during chemotaxis and calcium flux responses. *J Exp Med* 33: 1328–1333

93 Garzon R, Soriano SF, Rodriguez-Frade JM, Gomez L, Martin de Ana A, Sanchez-Gomez M, Martinez-A C, Mellado M (2004) CXCR4-mediated suppressor of cytokine signaling upregulation inactivates growth hormone function. *J Biol Chem* 279: 44460–44466

94 Ahr B, Denizot M, Robert-Hebmann V, Brelot A, Virad-Piechaczyk M (2005) Identification of the cytoplasmic domains of CXCR4 involved in Jak2 and Stat3 phosphorylation. *J Biol Chem* M408481200

95 Alexander WS (2002) Suppressors of cytokine signaling (SOCS) in the immune system. *Nat Rev Immunol* 2: 1–7

Chemokines in leukocyte transendothelial migration

Lixin Liu and Paul Kubes

Immunology Research Group, Department of Physiology and Biophysics, University of Calgary, 3330 Hospital Drive N.W., Calgary, Alberta T2N 4N1, Canada

Introduction

The recruitment of leukocytes from the blood stream to the site of infection or injury is of key importance in inflammation. The consequences of this recruitment can be the elimination of the invading pathogen but can also lead to inappropriate dysfunction. At inflammatory sites in the post-capillary venules of tissues, leukocyte recruitment involves complex interactions between leukocytes and endothelial cells characterized as firstly the tethering and rolling of leukocytes along the endothelium followed by leukocyte activation and firm adhesion to the endothelium, and then the migration of adherent leukocytes across the endothelium (diapedesis). Finally the emigrated leukocytes leave the vicinity of the venule and migrate toward the site of infection or injury guided by a gradient of one or more chemoattractants emanating from the afflicted site (chemotaxis). According to the currently accepted paradigm, the rolling is mediated by L-, P-, E-selectins and in some cases by α4 integrins, the adhesion is mediated by the activated α4 and β2 integrins, and the transmigration and subsequent chemotaxis in the tissues involves sophisticated cellular surface interactions and multiple signaling events among cell adhesion molecules, chemotactic signals and intracellular signaling pathways. Research advances very rapidly in this and related fields, and for more information on specific topics, readers are referred to a number of comprehensive reviews on the adhesion molecules on leukocytes and endothelial cells [1, 2], on the distribution of chemokines and chemokine receptors and their role in leukocyte migration [3–5], and on the role of cell adhesion molecules and cellular signaling mechanisms in leukocyte transendothelial migration [6–9]. Figure 1 gives a schematic summary of leukocyte transendothelial migration process in most tissues as exemplified by the inflammatory response in mesentery and cremaster muscle. However, in some organs including the lung, liver, and brain, the mechanisms can be distinct from this paradigm. Some reviews have been published highlighting the organ-specific mechanisms of leukocyte recruitment [6, 10, 11].

Chemokine Biology – Basic Research and Clinical Application, Volume I
edited by Bernhard Moser, Gordon L. Letts and Kuldeep Neote

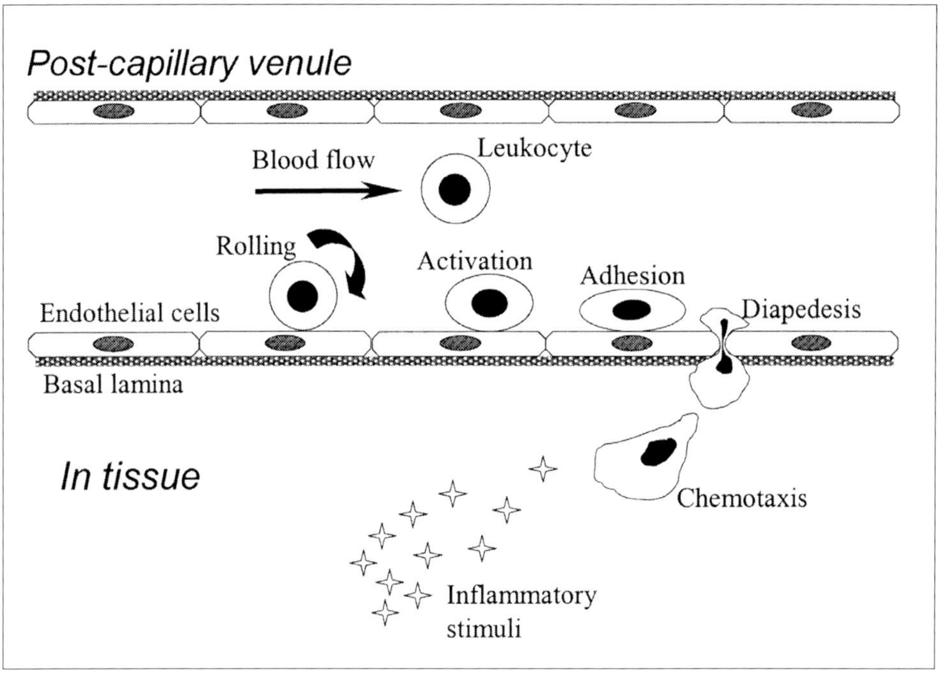

Figure 1
The scheme of leukocyte transendothelial migration process which occurs in most inflamed tissues in the body.

Under physiologic conditions, L-selectin, P-selectin glycoprotein ligand-1 and other molecules are constitutively expressed on the leukocyte surface. By contrast, P- and E-selectins are generally expressed on the lumenal surface of endothelial cells following appropriate activation. There may be some exception to this rule; P- and even E-selectin are constitutively expressed in skin and perhaps a few other organs. This allows leukocytes to roll on the activated endothelial cells. The integrins found primarily on leukocytes are normally in a low adhesive state. When activated, the low adhesive integrins can rapidly be induced into a high adhesive state and mediate binding with molecules of the immunoglobulin superfamily adhesion molecules (for example, intercellular adhesion molecule [ICAM]-1). This latter step results in the arrest of leukocytes on the lumenal surface of endothelium. The activation signal is thought to emanate mainly from endothelial cell surface-bound chemokines and sometimes from other chemoattractants, such as platelet-activating factor (PAF) and leukotriene (LT) B4. Chemokines are also important for leukocyte transmigration across the endothelium and subsequent chemotaxis in the tissues [12, 13].

Chemokines are a family of chemotactic cytokines that are secreted or membrane-bound, structurally related proteins of 67-127 amino acid peptides. There are about 50 chemokines in humans [3, 4, 14, 15], which fall in four subfamilies: CXC (α), CC (β), C (γ), and CX3C (δ) according to the number and location of the cysteine residues in the amino terminal end. Chemokines transmit the signals to the cells via binding to chemokine receptors which are all seven-transmembrane G-protein coupled receptors similar to cell surface receptors for other chemoattractants [3, 4, 14, 15]. Here we focus on the role for chemokines in leukocyte transendothelial migration and the contributions of selectins and signaling mechanisms in this process.

Chemokines trigger leukocyte adhesion to and transmigration across endothelium

Numerous studies have confirmed the multi-step leukocyte recruitment paradigm during inflammation. By using intravital microscopy, a powerful technique by which the leukocyte recruitment in tissues can be directly visualized and quantified, it has been established that this is a sequential process which occurs in the post-capillary venules in most inflamed tissues. Leukocytes initially tether and roll along the endothelium and then the rolling leukocytes adhere to the endothelial cells before transendothelial migration can happen. The selectins or α4 integrins tether cells to the endothelium. This localizes the cells to the endothelial surface, making it possible for leukocytes to sense chemokines presented by the inflamed microvasculature. These chemokines are either produced locally or reach the luminal site of blood vessels after transcytosis [16, 17]. Under flow conditions, chemokines have been shown to initiate leukocyte adhesion when they are co-immobilized with a selectin ligand and an integrin ligand or when the chemokines are immobilized on the surface of endothelial cells [18, 19].

Current wisdom suggests that chemokines must be immobilized to trigger rolling leukocytes to adhere. If chemokines remain soluble in the blood stream, they are washed away by the flow. The importance of endothelial cell-bound chemokines in bringing the rolling leukocytes to arrest was demonstrated by Weber et al. [20]. Their study showed that upon cytokine stimulation, endothelial cells produce chemokines of both endothelial-bound Gro-α (CXCL1) and soluble MCP-1 (CCL2). Under flow, MCP-1 that enters the vessel lumen is washed away by the fluid, but Gro-α remains on the endothelial cell surface. Therefore only Gro-α can bind to its receptor CXCR2 on monocytes. This activates monocytes and mediates the monocyte adhesion to endothelial cells. Although the soluble MCP-1 is unable to mediate monocyte adhesion, it is released in a manner (presumably abluminally) that allows the subsequent transendothelial migration [20, 21]. These studies confirmed that under physiologic flow condition, in order to trigger rolling leukocytes

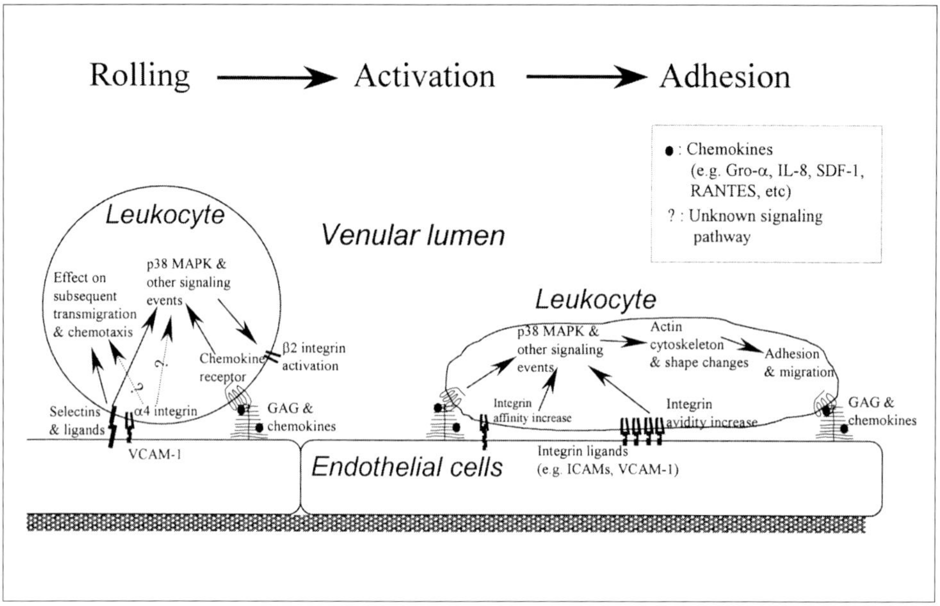

Figure 2
The summary of the signaling events that are related to leukocyte transendothelial migration in the cascade of leukocyte rolling, activation, and adhesion to inflamed endothelium.

to effectively adhere to endothelium, chemokines produced in the inflamed tissues must be immobilized on the surface of endothelial cells. Soluble chemokines are unlikely to be able to trigger this adhesion but can form a gradient to induce transendothelial migration.

Many chemokines produced and secreted in the inflammatory sites are immobilized on the endothelial cell surface via binding to glycosaminoglycans (GAGs), in particular heparan sulfate proteoglycans [17, 22]. Using electron microscopy, Middleton et al. [23] found that after chemokines interleukin (IL)-8 (CXCL8) and RANTES (CCL5) were injected into the skin, these chemokines were first found bound to GAGs at the abluminal surface of endothelium, then internalized into endothelial plasmalemmal vesicles and transported transcellularly on to the lumenal surface where the chemokines were presented to the rolling leukocytes. Recent studies found that a number of chemokines such as IL-8 (CXCL8), PF4 (CXCL4) and SDF-1 (CXCL12) can bind GAGs via the chemokine's C-terminal region [24–27]. For these chemokines, the GAG-binding domain was found to be spatially apart from the residues for binding and interacting with the chemokine receptors on leukocytes. This makes it possible that chemokines can activate the rolling leukocytes to adhere while binding with the GAGs.

Many chemokines have been shown to be important in the activation of integrins [28–30]. The integrins can rapidly undergo two different and dynamic ways of functional activation and allow binding to integrin ligands. One way involves the increase in integrin affinity by changing the three-dimensional conformation that leads to high affinity binding to the luminal side of blood vessels. The alternative way is the lateral mobility of integrins to a restricted area (also called clustering) to increase the avidity for the surface ligands [28–30]. Chemokines have been shown to be able to trigger both ways of integrin activation to support integrin-mediated adhesion of leukocytes to the ligands. The importance of chemokine-induced activation of integrin-mediated leukocyte adhesion was modeled in *in vitro* flow chamber systems which mimic the shear conditions seen under flow *in vivo*. In this system, leukocytes were allowed to flow under physiological shear conditions over cultured monolayers of endothelial cells which were activated by proinflammatory cytokines that stimulate endogenous chemokine and adhesion ligand production. Using this system, it was shown that many immobilized chemokines can trigger integrin-mediated leukocyte adhesion to endothelial cells and induce transendothelial migration [28–31]. Figure 2 summarizes the signaling events in both leukocytes and endothelial cells that are related to leukocyte transendothelial migration in the cascade of leukocyte rolling, activation, and adhesion to inflamed endothelial cells.

Chemokine-induced transendothelial migration requires engagement of selectins

The first way by which selectins may contribute to chemokine-induced leukocyte adhesion and subsequent emigration is to increase the length of time a cell interacts with a particular area of endothelium. This, for instance, could be achieved by a reduction in the rolling velocity. Interestingly, some inflammatory mediators (LTC_4, tumor necrosis factor (TNF)) but not all (histamine, H_2O_2) cause a down-modulation of the rolling velocity without necessarily inducing firm adhesion. However, this attenuated rolling behavior would then facilitate firm adhesion. There are a number of mechanisms by which slow rolling may occur, including a simple increase in the density of selectins and their ligands on leukocytes and endothelial cells.

The physiologic importance of slow rolling for chemokine function was demonstrated by a number of groups. Kanwar and colleagues [32] demonstrated that neither LTC_4 nor histamine induced adhesion but only LTC_4 induced slow rolling. Addition of low concentrations of proadhesive molecules (PAF, IL-8) induced adhesion only in those cells that were exposed to LTC_4. Only at much higher concentrations of PAF was adhesion observed with histamine which is known to induce P-selectin expression on the endothelial cells. Ley and colleagues [33] made similar observations by inducing slow rolling with TNF and then demonstrating that the slow rolling was dependent on E-selectin. When this molecule was inhibited, cells rolled faster and were less apt to respond to a local chemokine stimulus and less likely to adhere. An alternative explanation could be that the slow rolling was a result of the engagement of a significant number of selectin ligands (due to increased selectin density) which would cause signaling and subsequent predisposition for adhesion within rolling leukocytes. Also, low-level integrin activation has been shown to induce slow rolling [34].

The concept of signaling through selectins has been studied. Although there is little evidence of rapid physiologic changes following P-selectin cross-linking, there is good evidence that E-selectin can transmit signals to prepare cells for adhesion and transmigration. Using transfected L cells expressing human E-selectin and ICAM-1 in a parallel plate flow chamber assay, Simon and colleagues demonstrated that neutrophil tethering and rolling on E-selectin under flow conditions activate β2 integrins LFA-1 and Mac-1 to bind to the ligand ICAM-1 and that this E-selectin-mediated rolling transduces signals via mitogen-activated protein kinase (MAPK) to induce neutrophil arrest on ICAM-1 [35]. This signaling event was further demonstrated in neutrophil recruitment on endothelial cells and that the E-selectin engagement stimulates both the clustering and high affinity of β2 integrins and mediates the binding of neutrophils to β2 integrin ligands via p38 and p42/44 MAPK signaling [36]. Although these data suggest that E-selectin and β2 integrins can function independent of chemokines, the physiological role is likely to enhance chemokine-induced integrin activation.

Chemokines do not seem to act alone but interact with other factors such as shear in a coordinated fashion to induce efficient leukocyte transendothelial migration [28, 31]. L-selectin has long been known to mediate the initial leukocyte tethering and rolling along the inflamed endothelium in peripheral tissues. Although L-selectin may play some role in rolling in the periphery, there is a growing body of evidence to suggest that this molecule can enhance chemokine function and have a large impact on the subsequent leukocyte transmigration process. Earlier *in vitro* studies revealed that cross-linking L-selectin upregulated the β2 integrin Mac-1 and increased the binding to its ligand in the presence of chemokines [37]. Using a laminar flow chamber assay, Simon at al. [38] showed that in the presence of lipopolysaccharide (LPS)-induced endothelial chemokine production, stimulation of L-selectin via cross-linking dramatically increased the capacity of neutrophils to firmly attach and spread on endothelium, and migrate across the endothelial cell monolayer. Tsang and colleagues [39] also demonstrated that cross-linking of L-selectin potentiated IL-8-stimulated leukocyte shape change and synergistically enhanced β2 integrin-mediated neutrophil adhesion to and transmigration across cytokine-stimulated endothelial cells.

In vivo studies also suggest a role for L-selectin in enhancing chemokine functions. Using L-selectin-deficient mice, Hickey et al. [40] examined the role of L-selectin in chemokine-induced neutrophil transendothelial migration and chemotaxis in an acute inflammation model. In this model, an agarose gel containing mouse CXC chemokine keratinocyte-derived chemokine (KC/CXCL1) was placed

350 μm from a post-capillary venule in cremaster muscle to induce neutrophil transmigration and chemotaxis toward the slow-releasing chemokine KC. This study found no inhibition of leukocyte rolling or adhesion in L-selectin-deficient mice. However, there was a 60% reduction of neutrophil emigration and for the remaining 40% of cells that did emigrate across the endothelium, the cells remained closely associated with the venules rather than chemotaxing toward the KC-containing gel. The importance of these results were further underscored by a report by Grewal et al. [41] in experimental autoimmune encephalomyelitis, an animal model of multiple sclerosis. In this study, L-selectin was found to be essential for the mice to develop antigen-induced experimental autoimmune encephalomyelitis and to mediate myelin damage. Upon closer examination of the brains of these mice it became evident that in the L-selectin-deficient mice, leukocytes crossed the blood brain barrier but were unable to chemotax away from the vasculature [41].

Chemokine-induced transendothelial migration and p38 MAPK signaling

Chemokine-induced leukocyte transendothelial migration is dependent upon a number of signaling pathways within the leukocytes as well as the endothelium. There has been ample evidence that engagement of L-selectin induces activation of several signal transduction pathways including activation of p38 MAPK. To explore the role of p38 MAPK in leukocyte recruitment *in vivo*, Cara and colleagues used p38 inhibitors and examined chemokine KC-induced leukocyte recruitment in mice [42]. It was found that p38 MAPK inhibitors at concentrations previously demonstrated to be anti-inflammatory had no effect in leukocyte rolling along the endothelial surface, or adhesion to endothelium, but dramatically inhibited leukocyte transendothelial migration. The leukocyte chemotaxis in the cremaster muscle tissue was also inhibited by p38 MAPK inhibition. This study suggested that the p38 MAPK downstream of L-selectin may be important in chemokine KC-induced leukocyte emigration and chemotaxis. Because it has been shown that *in vitro* a p38 MAPK inhibitor eliminated chemokine-induced murine neutrophil chemotaxis toward CXC chemokines KC and MIP-2 [43], it is thus unclear whether the contribution of p38 MAPK in neutrophil transmigration and chemotaxis *in vivo* is due to the downstream effect of L-selectin or the downstream effect of chemokine receptors. However, others have reported no effect of p38 MAPK inhibition on CXC chemokine-induced chemotaxis raising the possibility that p38 MAPK inhibitors could either be targeting the L-selectin-dependent aspect of the emigration process or alternatively, since the whole mouse was being treated with p38 MAPK inhibitors [42], even non-hematopoietic cells, such as endothelial cells, could have been affected (discussed in the next section).

Activation of endothelial p38 MAPK is necessary for leukocyte transmigration. For example, the endothelial p38 MAPK signaling pathway is activated by the interactions between integrins and their ligands (e.g., ICAM-1 and VCAM-1) that mediate leukocyte firm arrest on endothelium. Wang and Doerschuk demonstrated that cross-linking ICAM-1 on endothelial cells which mimics leukocyte binding to endothelium, induced phosphorylation of p38 MAPK and increased downstream activity [44]. They showed that the activation of p38 MAPK was responsible for the activation of one of the downstream effectors heat shock protein 27 which is involved in F-actin polymerization in endothelial cells. Studies from that group recently revealed that kinases up-stream of p38 MAPK such as MKK3 and MKK6 are also required for this response, and that inhibition of p38α (one of the isoforms of p38 MAPK) attenuated ICAM-1-dependent endothelial cytoskeletal changes and attenuated neutrophil migration to the endothelial cell borders [45].

VCAM-1, another member of immunoglobulin superfamily adhesion molecules is the ligand for α4 integrins. Using antibody-mediated cross-linking, van Wetering et al. found that engagement of VCAM-1 on interleukin-1-activated endothelial cells induced endothelial cell actin stress fiber formation, contractility, activation of p38 MAPK and formation of endothelial cellular gaps [46]. These researchers further demonstrated that (1) inhibition of p38 MAPK largely prevented the effects of VCAM-1 engagement on endothelial F-actin stress fiber induction and endothelial cell-cell gap formation, (2) the phosphorylation of p38 MAPK by VCAM-1 engagement was downstream of the signaling of Rac, a member of Rho small GTPase family, and (3) inhibition of Rac function significantly attenuated leukocyte transendothelial migration. These studies suggested that leukocyte adhesion to endothelial cells can activate both p38 MAPK and the downstream cytoskeletal changes that regulate the transendothelial migration of leukocytes.

Leukocyte-specific protein 1 (LSP1) has been shown to be one of the major substrates of MAPK-activated protein kinase-2 which is directly downstream of p38 MAPK [47]. LSP1 is an intracellular F-actin-binding and Ca^{2+}-binding protein and was initially found to be expressed only in leukocytes [48–50]. Therefore, using LSP1-deficient mice, it was not surprising that LSP1 was found to be involved in chemokine-induced leukocyte emigration *in vivo* [51] and neutrophil chemotaxis *in vitro* [52]. What was more unexpected was that LSP1 was also expressed in mouse and human endothelial cells [53]. By using RT-PCR, western blotting and immunofluorescent microscopy, it was demonstrated that both murine primary microvascular endothelial cells and human umbilical vein endothelial cells expressed LSP1. Endothelial LSP1 regulated chemokine-induced leukocyte transendothelial migration by playing an important role in endothelial cells probably through the regulation of cytoskeletal change-related endothelial cell retraction [53]. Therefore, LSP1 in endothelial cells is also likely an important player in chemokine-induced leukocyte transendothelial migration.

After transendothelial migration, emigrated leukocytes must begin to orient themselves according to the local chemokine gradient for directional movement to infections or tissue injuries where the concentration of inflammatory chemokines is

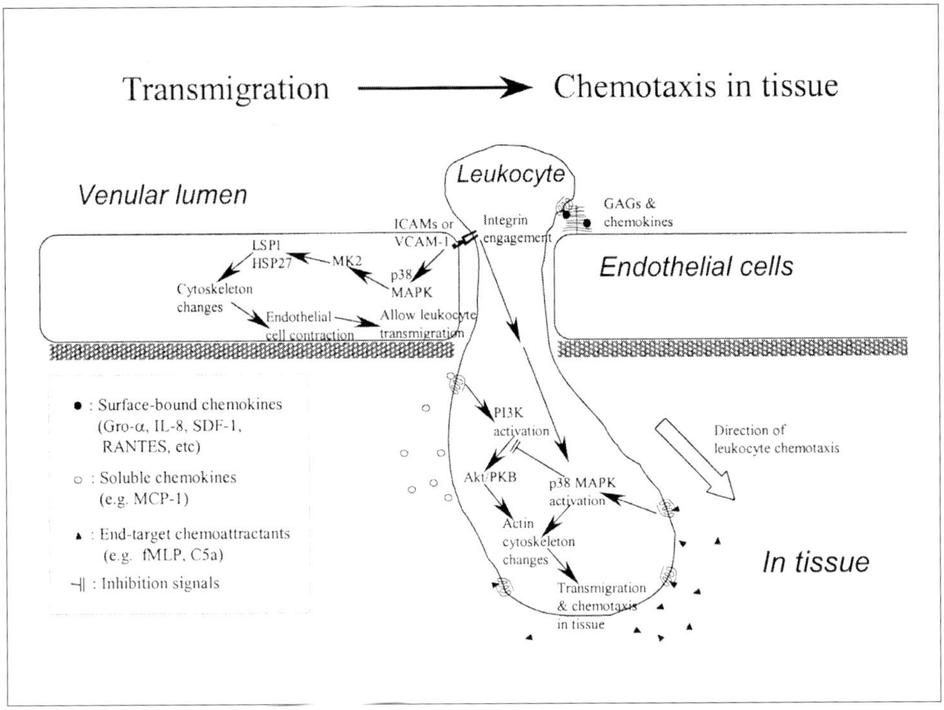

Figure 3
The signaling pathways in leukocytes and endothelial cells during chemokine-induced leukocyte transendothelial migration and subsequent chemotaxis toward end-target chemoattractants in inflamed tissues. MK2, MAPK-activated protein kinase-2. HSP27, heat shock protein 27. PI3K, phosphoinositide 3-kinase. PKB, protein kinase B, also known as Akt

highest (for a comprehensive review, see reference [4]). Clearly these cells need to ignore the endothelium-associated chemokines in order to be able to respond to the inflammatory chemotactic gradients in the tissue. It was found *in vitro* that neutrophils will selectively migrate toward end-target chemoattractants (which are bacterial products or activated complement fragments produced exclusively at the site of infection or tissue injury and chemotactic for leukocytes, such as fMLP or C5a) and ignore or override the presence of chemokines such as IL-8 [54, 55]. Heit et al. showed that fMLP or C5a activates leukocyte p38 MAPK signaling pathway and provides an inhibitory signal for other signaling pathways (such as phosphoinositide 3-kinase and the downstream Akt/PKB activation) normally induced by chemokines [55]. Thus neutrophils can differentiate signaling events and migrate preferentially toward the end-target chemoattractants produced during infection or injury in the tissue. Figure 3 shows a brief summary of current understanding of the signaling pathways in both leukocytes and endothelial cells in chemokine-induced transendothelial migration and subsequent leukocyte chemotaxis toward end-target chemoattractants in inflamed tissues.

Conclusions

Chemokines function at all stages of leukocyte transendothelial migration. However, chemokines do not work alone. Selectins enhance chemokine-induced leukocyte transendothelial migration. Activation of p38 MAPK plays an important role in chemokine-induced transmigration. Unraveling the mechanisms of leukocyte transendothelial migration and the signaling pathways involved is now a major area of interest. Interactions between chemokines, adhesion molecules, the cytoskeleton, signaling kinases and other signaling factors need further exploration to provide new clues for novel therapies for the treatment of inflammatory diseases.

Acknowledgements
L. Liu is supported by a research fellowship from Alberta Heritage Foundation for Medical Research (AHFMR). P. Kubes is an AHFMR scientist and a Canada Research Chair.

References

1 Carlos TM, Harlan JM (1994) Leukocyte-endothelial adhesion molecules. *Blood* 84: 2068–2101
2 Aplin AE, Howe A, Alahari SK, Juliano RL (1998) Signal transduction and signal modulation by cell adhesion receptors: the role of integrins, cadherins, immunoglobulin-cell adhesion molecules, and selectins. *Pharmacol Rev* 50: 197–263
3 Olson TS, Ley K (2002) Chemokines and chemokine receptors in leukocyte trafficking. *Am J Physiol Regul Integr Comp Physiol* 283: R7–R28
4 Moser B, Wolf M, Walz A, Loetscher P (2004) Chemokines: multiple levels of leukocyte migration control. *Trends Immunol* 25: 75–84
5 Rot A, von Andrian UH (2004) Chemokines in innate and adaptive host defense: basic chemokinese grammar for immune cells. *Annu Rev Immunol* 22: 891–928
6 Liu L, Kubes P (2003) Molecular mechanisms of leukocyte recruitment: organ-specific mechanisms of action. *Thromb Haemost* 89: 213–220
7 Liu Y, Shaw SK, Ma S, Yang L, Luscinskas FW, Parkos CA (2004) Regulation of leukocyte transmigration: cell surface interactions and signaling events. *J Immunol* 172: 7–13

8 van Buul JD, Hordijk PL (2004) Signaling in leukocyte transendothelial migration. *Arterioscler Thromb Vasc Biol* 24: 824–833

9 Muller WA (2003) Leukocyte-endothelial-cell interactions in leukocyte transmigration and the inflammatory response. *Trends Immunol* 24: 326–333

10 Doerschuk CM (2001) Mechanisms of leukocyte sequestration in inflamed lungs. *Microcirculation* 8: 71–88

11 Mizgerd JP (2002) Molecular mechanisms of neutrophil recruitment elicited by bacteria in the lungs. *Semin Immunol* 14: 123–132

12 Nourshargh S, Marelli-Berg FM (2005) Transmigration through venular walls: a key regulator of leukocyte phenotype and function. *Trends Immunol* 26: 157–165

13 Ebnet K, Vestweber D (1999) Molecular mechanisms that control leukocyte extravasation: the selectins and the chemokines. *Histochem Cell Biol* 112: 1–23

14 Moser B, Willimann K (2004) Chemokines: role in inflammation and immune surveillance. *Ann Rheum Dis* 63: ii84–ii89

15 Ono SJ, Nakamura T, Miyazaki D, Ohbayashi M, Dawson M, Toda M (2003) Chemokines: roles in leukocyte development, trafficking, and effector function. *J Allergy Clin Immunol* 111: 1185–1199

16 Rot A, Hub E, Middleton J, Pons F, Rabeck C, Thierer K, Wintle J, Wolff B, Zsak M, Dukor P (1996) Some aspects of IL-8 pathophysiology. III: Chemokine interaction with endothelial cells. *J Leukoc Biol* 59: 39–44

17 Middleton J, Patterson AM, Gardner L, Schmutz C, Ashton BA (2002) Leukocyte extravasation: chemokine transport and presentation by the endothelium. Blood 100: 3853–3860

18 Campbell JJ, Hedrick J, Zlotnik A, Siani MA, Thompson DA, Butcher EC (1998) Chemokines and the arrest of lymphocytes rolling under flow conditions. *Science* 279: 381–384

19 Rainger GE, Fisher AC, Nash GB (1997) Endothelial-borne platelet-activating factor and interleukin-8 rapidly immobilize rolling neutrophils. *Am J Physiol* 272: H114–H122

20 Weber KS, von Hundelshausen P, Clark-Lewis I, Weber PC, Weber C (1999) Differential immobilization and hierarchical involvement of chemokines in monocyte arrest and transmigration on inflamed endothelium in shear flow. *Eur J Immunol* 29: 700–712

21 Randolph GJ, Furie MB (1995) A soluble gradient of endogenous monocyte chemoattractant protein-1 promotes the transendothelial migration of monocytes *in vitro*. *J Immunol* 155: 3610–3618

22 Weber C (2003) Novel mechanistic concepts for the control of leukocyte transmigration: specialization of integrins, chemokines, and junctional molecules. *J Mol Med* 81: 4–19

23 Middleton J, Neil S, Wintle J, Clark-Lewis I, Moore H, Lam C, Auer M, Hub E, Rot A (1997) Transcytosis and surface presentation of IL-8 by venular endothelial cells. *Cell* 91: 385–395

24 Maione TE, Gray GS, Hunt AJ, Sharpe RJ (1991) Inhibition of tumor growth in mice by an analogue of platelet factor 4 that lacks affinity for heparin and retains potent angiostatic activity. *Cancer Res* 51: 2077–2083
25 Webb LM, Ehrengruber MU, Clark-Lewis I, Baggiolini M, Rot A (1993) Binding to heparan sulfate or heparin enhances neutrophil responses to interleukin 8. *Proc Natl Acad Sci USA* 90: 7158–7162
26 Kuschert GS, Hoogewerf AJ, Proudfoot AE, Chung CW, Cooke RM, Hubbard RE, Wells TN, Sanderson PN (1998) Identification of a glycosaminoglycan binding surface on human interleukin-8. *Biochemistry* 37: 11193–11201
27 Luo J, Luo Z, Zhou N, Hall JW, Huang Z (1999) Attachment of C-terminus of SDF-1 enhances the biological activity of its N-terminal peptide. *Biochem Biophys Res Commun* 264: 42–47
28 Cinamon G, Grabovsky V, Winter E, Franitza S, Feigelson S, Shamri R, Dwir O, Alon R (2001) Novel chemokine functions in lymphocyte migration through vascular endothelium under shear flow. *J Leukoc Biol* 69: 860–866
29 Johnston B, Butcher EC (2002) Chemokines in rapid leukocyte adhesion triggering and migration. *Semin Immunol* 14: 83–92
30 Laudanna C, Kim JY, Constantin G, Butcher E (2002) Rapid leukocyte integrin activation by chemokines. *Immunol Rev* 186: 37–46
31 Cuvelier SL, Patel KD (2001) Shear-dependent eosinophil transmigration on interleukin 4- stimulated endothelial cells: a role for endothelium-associated eotaxin-3. *J Exp Med* 194: 1699–1709
32 Kanwar S, Johnston B, Kubes P (1995) Leukotriene C_4/D_4 induces P-selectin and sialyl Lewis^x- dependent alterations in leukocyte kinetics *in vivo*. *Circ Res* 77: 879–887
33 Ley K, Allietta M, Bullard DC, Morgan S (1998) Importance of E-selectin for firm leukocyte adhesion *in vivo*. *Circ Res* 83: 287–294
34 Kubes P, Kerfoot SM (2001) Leukocyte recruitment in the microcirculation: the rolling paradigm revisited. *News Physiol Sci* 16: 76–80
35 Simon SI, Hu Y, Vestweber D, Smith CW (2000) Neutrophil tethering on E-selectin activates β2 integrin binding to ICAM-1 through a mitogen-activated protein kinase signal transduction pathway. *J Immunol* 164: 4348–4358
36 Green CE, Pearson DN, Camphausen RT, Staunton DE, Simon SI (2004) Shear-dependent capping of L-selectin and P-selectin glycoprotein ligand 1 by E-selectin signals activation of high-avidity β_2-integrin on neutrophils. *J Immunol* 172: 7780–7790
37 Crockett-Torabi E, Sulenbarger B, Smith CW, Fantone JC (1995) Activation of human neutrophils through L-selectin and Mac-1 molecules. *J Immunol* 154: 2291–2302
38 Simon SI, Burns AR, Taylor AD, Gopalan PK, Lynam EB, Sklar LA, Smith CW (1995) L-selectin (CD62L) cross-linking signals neutrophil adhesive functions via the Mac-1 (CD11b/CD18) β_2-integrin. *J Immunol* 155: 1502–1514
39 Tsang YTM, Neelamegham S, Hu Y, Berg EL, Burns AR, Smith CW, Simon SI (1997)

Synergy between L-selectin signaling and chemotactic activation during neutrophil adhesion and transmigration. *J Immunol* 159: 4566–4577
40 Hickey MJ, Forster M, Mitchell D, Kaur J, De Caigny C, Kubes P (2000) L-selectin facilitates emigration and extravascular locomotion of leukocytes during acute inflammatory responses *in vivo*. *J Immunol* 165: 7164–7170
41 Grewal IS, Foellmer HG, Grewal KD, Wang H, Lee WP, Tumas D, Janeway CA Jr, Flavell RA (2001) CD62L is required on effector cells for local interactions in the CNS to cause myelin damage in experimental allergic encephalomyelitis. *Immunity* 14: 291–302
42 Cara DC, Kaur J, Forster M, McCafferty DM, Kubes P (2001) Role of p38 mitogen-activated protein kinase in chemokine-induced emigration and chemotaxis *in vivo*. *J Immunol* 167: 6552–6558
43 Nick JA, Young SK, Brown KK, Avdi NJ, Arndt PG, Suratt BT, Janes MS, Henson PM, Worthen GS (2000) Role of p38 mitogen-activated protein kinase in a murine model of pulmonary inflammation. *J Immunol* 164: 2151–2159
44 Wang Q, Doerschuk CM (2001) The p38 mitogen-activated protein kinase mediates cytoskeletal remodeling in pulmonary microvascular endothelial cells upon intracellular adhesion molecule-1 ligation. *J Immunol* 166: 6877–6884
45 Wang Q, Yerukhimovich M, Gaarde WA, Popoff IJ, Doerschuk CM (2005) MKK3 and -6-dependent activation of p38α MAP kinase is required for cytoskeletal changes in pulmonary microvascular endothelial cells induced by ICAM-1 ligation. *Am J Physiol Lung Cell Mol Physiol* 288: L359–L369
46 Van Wetering S, van den Berk N, van Buul JD, Mul FP, Lommerse I, Mous R, ten Klooster JP, Zwaginga JJ, Hordijk PL (2003) VCAM-1-mediated Rac signaling controls endothelial cell–cell contacts and leukocyte transmigration. *Am J Physiol Cell Physiol* 285: C343–C352
47 Huang CK, Zhan L, Ai Y, Jongstra J (1997) LSP1 is the major substrate for mitogen-activated protein kinase-activated protein kinase 2 in human neutrophils. *J Biol Chem* 272: 17–19
48 Klein DP, Galea S, Jongstra J (1990) The lymphocyte-specific protein LSP1 is associated with the cytoskeleton and co-caps with membrane IgM. *J Immunol* 145: 2967–2973
49 Klein DP, Jongstra-Bilen J, Ogryzlo K, Chong R, Jongstra J (1989) Lymphocyte-specific Ca^{2+}-binding protein LSP1 is associated with the cytoplasmic face of the plasma membrane. *Mol Cell Biol* 9: 3043–3048
50 Jongstra-Bilen J, Janmey PA, Hartwig JH, Galea S, Jongstra J (1992) The lymphocyte-specific protein LSP1 binds to F-actin and to the cytoskeleton through its COOH-terminal basic domain. *J Cell Biol* 118: 1443–1453
51 Jongstra-Bilen J, Misener VL, Wang C, Ginzberg H, Auerbach A, Joyner AL, Downey GP, Jongstra J (2000) LSP1 modulates leukocyte populations in resting and inflamed peritoneum. *Blood* 96: 1827–1835
52 Hannigan M, Zhan L, AiY, Huang CK (2001) Leukocyte-specific gene 1 protein

(LSP1) is involved in chemokine KC-activated cytoskeletal reorganization in murine neutrophils *in vitro*. *J Leukoc Biol* 69: 497–504
53 Liu L, Cara DC, Kaur J, Raharjo E, Mullaly SC, Jongstra-Bilen J, Jongstra J, Kubes P (2005) LSP1 is an endothelial gatekeeper of leukocyte transendothelial migration. *J Exp Med* 201: 409–418
54 Campbell JJ, Foxman EF, Butcher EC (1997) Chemoattractant receptor cross talk as a regulatory mechanism in leukocyte adhesion and migration. *Eur J Immunol* 27: 2571–2578
55 Heit B, Tavener S, Raharjo E, Kubes P (2002) An intracellular signaling hierarchy determines direction of migration in opposing chemotactic gradients. *J Cell Biol* 159: 91–102

Natural chemokine antagonism and synergism

Mariagrazia Uguccioni[1] *and Basil O. Gerber*[2]

[1]Institute for Research in Biomedicine, CH-6500 Bellinzona, Switzerland; [2]Division of Allergology, Dept. of Rheumatology & Clinical Immunology/Allergology, Inselspital, CH-3011 Bern, Switzerland

Introduction

It is now generally accepted that leukocyte trafficking in homeostasis as well as in pathology is largely determined by the more than 40 chemokines that are produced constitutively or upon specific induction in virtually all tissues of the human body, in combination with the expression of almost 20 target receptors on all leukocyte subsets and on many tissue cells. Although much remains to be discovered, the receptor specificities of most chemokines, expression patterns of chemokine receptors, and the resulting immunologic activities are often known in intricate detail from numerous *in vitro* and *in vivo* studies.

While we thus understand well the effects of chemokines one by one, much less is known of the potential consequences of multiple and concomitant chemokine expression on leukocyte migration and function, even though numerous *in situ* experiments clearly document the simultaneous expression of several or many chemokines at diverse target sites of leukocyte trafficking and homing. Evidence from other and our own groups has recently revealed the existence of additional modulatory mechanisms that apply under conditions of multiple and concomitant chemokine expression. Here, we summarise our current knowledge of the negative or positive influence that such a chemokine "milieu" can exert by natural chemokine antagonism and synergism.

Natural chemokine antagonism

The term "natural antagonist" has become customary to designate endogenous, full-length chemokines that feature inhibitory activities distinct of and in addition to their agonistic properties. Strictly spoken, such natural antagonists also include chemokines that have acquired their inhibitory properties by protease modification,

Chemokine Biology – Basic Research and Clinical Application, Volume I
edited by Bernhard Moser, Gordon L. Letts and Kuldeep Neote

Table 1 - Summary of protease-modified chemokines with antagonistic activities

Chemokine	Modifying enzyme	Antagonist for
CCL2	MMP1/3 [10]	CCR2/3
CCL5	CD26 [3], not specified [4]	CCR1/3
CCL7	(MT1)-MMP, MMP-1/2/3/13[7, 10]	CCR2/3
CCL8	MMP1/3 [10], not specified [2]	CCR2/3
CCL11	CD26 [5]	CCR3
CCL13	MMP1/3 [10]	CCR2/3
CCL22	CD26 [6]	CCR4
CXCL9	CD26 [9]	CXCR3
CXCL10	CD26 [9]	CXCR3
CXCL11	CD26 [8, 9]	CXCR3
CXCL12	CD26 [11, 12]	CXCR4

as well as viral chemokine homologues with inhibitory potential. Viral chemokines and chemokine receptors will be discussed elsewhere in this volume and are thus not considered further. Here, we will focus on endogenous, human chemokines, and use the terms "protease-modified" or "native" for further distinction.

Protease-modified chemokines

While chemokines are very resistant to proteolytic degradation and inactivation in general, specific processing can occur in the N-terminal and C-terminal domains. Various enzymes, namely dipeptidyl peptidase IV (DPP-IV/CD26) and matrix metalloproteinases (MMPs), can process chemokines, thus generating completely inactive chemokines, chemokine antagonists, and chemokines with altered receptor selectivity or increased activity [1].

The 11 chemokines that are known to be converted to inhibitory chemokines by protease digestion are summarised in Table 1 [2–12], together with the converting enzymes and the six target receptors (compare to Tab. 3 in [13]). The fact that N-terminal protease digestion often produces inhibitory chemokines is compatible with the body of structural and structure-function studies (reviewed in [14, 15]), which collectively indicate the N-terminus of most chemokines as the receptor-activating domain, while the random-coiled N-loop distal of the first two conserved cysteines, together with residues situated in the third β-strand, form the receptor binding domain. This spatial separation allows the easy removal or truncation of the activation domain, resulting in a receptor-binding, "dominant negative" chemokine.

Table 2 - Summary of native chemokines with antagonistic activities, listed by chemokines (top half) and target receptors (bottom half)

Chemokine	Agonist for	Antagonist for
CCL4	CCR5	CCR1 [21]
CCL7	CCR1, CCR2, CCR3	CCR5 [17]
CCL11	CCR3	CCR2 [19, 20, 27, 28]
CCL18	not known	CCR3 [18, 22]
CCL24	CCR3	CCR2 [27]
CCL26	CCR3	CCR1 [26], CCR2 [24, 27, 28], CCR5 [26]
CXCL9	CXCR3	CCR3 [16]
CXCL10	CXCR3	CCR3 [16]
CXCL11	CXCR3	CCR3 [16, 22], CCR5 [25]

Receptor	Agonists	Antagonists
CCR1	CCL3/5/7/8/13/14/15/23	CCL4 [21], CCL26 [26]
CCR2	CCL2/7/8/13	CCL11 [19, 20, 27, 28], CCL24 [27], CCL26 [24]
CCR3	CCL5/7/8/11/13/24/26	CCL18 [18, 22], CXCL9/10 [16], CXCL11 [16, 22]
CCR5	CCL3/4/5/8	CCL7 [17], CCL26 [26], CXCL11 [25]

Native chemokines

The above two-site model of chemokine receptor binding and activation also implies that a native chemokine featuring a matching binding domain and a "mismatched" activation domain might act as an antagonist for a particular receptor just as well. In fact, a CCL11 hybrid with its N-terminus substituted by that of CXCL11 acted as an antagonist for CCR3, supporting this concept [16]. Altogether, nine native chemokines are currently known to have inhibitory activities apart from their previously known agonism. They are summarised in Table 2 [16–28], listed by chemokines as well as target receptors. Most show narrow antagonist specificity, inhibiting only one receptor. Notable exceptions are CXCL11 and CCL26, which are specific agonists for CXCR3 and CCR3, respectively, but inhibit two (CCR3 and CCR5) and three (CCR1, CCR2, and CCR5) receptors, respectively.

Mode of action

The current data suggest the notion that endogenous chemokines – be they in their native or protease-modified form – inhibit their target receptor by competitive antagonism[1], much as it is known for many other G protein coupled receptors. The action of CCL11 on CCR2 seems to be more complex, though. Initially described as an antagonist [19], which would make it a neutral (or possibly inverse) agonist in pharmacological terms, it was later reported to be a partial agonist [20, 27]. Different cellular backgrounds and differing receptor expression levels may account for these differences, again in analogy to other G protein coupled receptors. Interestingly, an unusual mechanism of active inhibition, involving receptor and mitogen-activated protein kinase (MAPK) activation, contributes to the observed antagonism [28].

In vivo relevance

To achieve their inhibitory effects, many endogenous antagonists require concentrations that far exceed those required for their agonistic actions in *in vitro* experiments. This has raised doubts if endogenous antagonists are produced in sufficient quantities to be of physiological relevance at all. However, N-terminally truncated, synthetic [29, 30] as well as protease-modified [7] chemokine antagonists have previously demonstrated their antagonistic potential *in vivo* in several rodent models. More recently, the native form of CXCL9, a somewhat modest CCR3 inhibitor *in vitro* [16], was found to be an efficient *in vivo* antagonist as well [31]. Interestingly, CXCL9 was unexpectedly identified together with other Th1-associated genes during a screen for "signature genes" of allergic airway inflammation in mice. CXCL9 inhibited IL-13- and chemokine-induced eosinophil migration to the lung and blood, as well as their functional responses. Notably, the inhibitory effects of CXCL9 were comparable to those seen in CCL11 or CCR3 gene-deleted mice, suggesting that natural antagonists may indeed exert a profound influence on the modulation of certain immune responses.

Natural chemokine synergism

An abundant number of publications describe various forms of synergism between different proinflammatory substances, cytokines, chemoattractants and chemo-

1 Here, we use the following definitions: a competitive antagonist progressively inhibits a response in the presence of a fixed agonist concentration. At full receptor occupancy, a partial agonist elicits a lower response than a full agonist, while neutral and inverse agonists do not induce any responses at all. Additionally, an inverse agonist also inhibits the constitutive activity of a receptor.

kines, involving many growth hormones, cytokine, Toll-like and G protein coupled receptors. Here, we will focus on synergistic combinations of chemokines and chemoattractants, which all act via the latter receptor family. It seems likely that two different mechanisms occur. On one hand, chemokine synergism may be due to intracellular priming events that are (probably) akin to those seen with proinflammatory substances, cytokines, and chemoattractants. On the other hand, chemokines seem to be capable of forming heteromeric complexes that are more active than the single chemokines or their homomeric complexes themselves, as discussed below.

Chemokine synergism by intracellular priming events

Regakine [32], a bovine chemokine with no known human orthologue to date, can specifically increase the activity of certain chemokines and chemoattractants such as CXCL6 [33], CXCL7 [34], CXCL8 [35], *N*-formylmethionylleucylphenylalanine (fMLP) [33], and complement factor 5a (C5a) [34]. The authors described a similar synergism for CXCL8 in the presence of CCL2, CCL7, CCL8, and CXCL12 [35]. Similar findings were obtained with haematopoietic stem/progenitor cells and combinations of C3a and CXCL12, where chemotaxis, metalloproteinase-9 secretion and cellular adhesion were all enhanced [36]. Another, reciprocal synergism modulates the responses of CXCR3 and CXCR4 to their agonists: CXCL12 primes the responsiveness of CXCR3$^+$, natural IFN-producing cells to CXCL9, CXCL10, and CXCL11 [37], while the reactivity of CXCR4$^+$ plasmacytoid dendritic cells to CXCL12 is similarly increased in the presence of CXCL9, CXCL10, and CXCL11 [38]. For two reasons, the authors of these reports suggested receptor-dependent priming as the most likely mechanism for the synergistic events: the expression of both receptors specific for the synergising components was required, and the structural differences between the chemokines and chemoattractants used make direct ligand interactions appear unlikely. The nature of the priming mechanism presumably causing the synergistic events remains to be determined for all of these systems, however. Interestingly, a recent report describes a novel kind of haptotactic chemorepulsion for CXCR3$^+$ plasmacytoid dendritic cells, which is cell-specific, independent of CXCR4-induced synergism, and inhibited by soluble CXCR3 agonists [39].

Chemokine synergism by heteromeric chemokine interactions

By nuclear magnetic resonance (NMR) and plasmon resonance-based Biacore analysis, two recent reports clearly demonstrate that CXCL4 and CXCL8 form heterodimers that were more active in haematopoiesis and chemotaxis assays than the respective chemokines on their own [40, 41]. CXCL4 and CXCL8 interact via their β-sheets, akin to how their homomeric complexes form [42]. Of note, these findings may furnish an explanation for synergistic effects observed in proliferation assays more than a decade earlier [43]. In another study, CXCL4 interacted with CCL5 in a heteromeric and synergistic way, increasing monocyte adherence on activated endothelial cells [44]. CCL5 mutated at position 26, a residue located in the first β-strand, is more prone to homomeric tetramer formation than the native form but was consequently refractory to synergism.

Synergism induced by heteromeric chemokine complexes may well be a widespread but nevertheless specific phenomenon, as documented by the large number (20 out of 25 tested) of chemokines that synergistically increased the action of CCL19 and CCL21 on CCR7 [45]. Apart from chemotaxis of CCR7 transfected cells, dendritic cells, and T and B cells, receptor internalisation and extracellular-regulated kinase (ERK) phosphorylation of transfectants were synergistically increased as well. Western blot and binding experiments again suggested the formation of heteromeric complexes as the cause of the observed synergism. At equal concentrations, a mixture of synergy-inducing chemokines was just as potent as any of the used chemokines alone at evoking synergism, suggesting that the effects were not just additive but truly synergistic. Similar synergy mechanisms enhance CCR4 responses towards CCL17 and CCL22 [46]. Interestingly, chimeric mutants between two chemokines with (CCL7) and without (CCL4) synergistic activity [46] imply that residues in the first β-strand mediate heteromeric association and synergism, much in analogy to the interaction between CXCL4 and CCL5 [44].

Mode of action

Taken together, the above reports suggest that synergism by heteromeric chemokine interactions may be a widespread phenomenon, positively regulating diverse chemokine activities such as chemotaxis, cellular adherence, receptor internalisation, and protein kinase phosphorylation. Interestingly, the available structure and structure-function data, albeit scarce to date, collectively implicate residues in the first β-strand as mediators of heteromeric association and synergism. It is thus tempting to speculate that heteromeric chemokine complexes may mimic those homomeric dimers that form via association of their β-sheets, featuring an interface composed of the first β-strands (see Fig. 1 and legend for further explanation). However, the molecular reasons as to why a heteromeric complex should be more active than a homomeric one remain at present completely obscure. Certainly, speculating that heteromeric chemokine association might promote receptor (hetero-) dimerisation, which was reported to increase receptor activities [47], would constitute an attractive hypothesis.

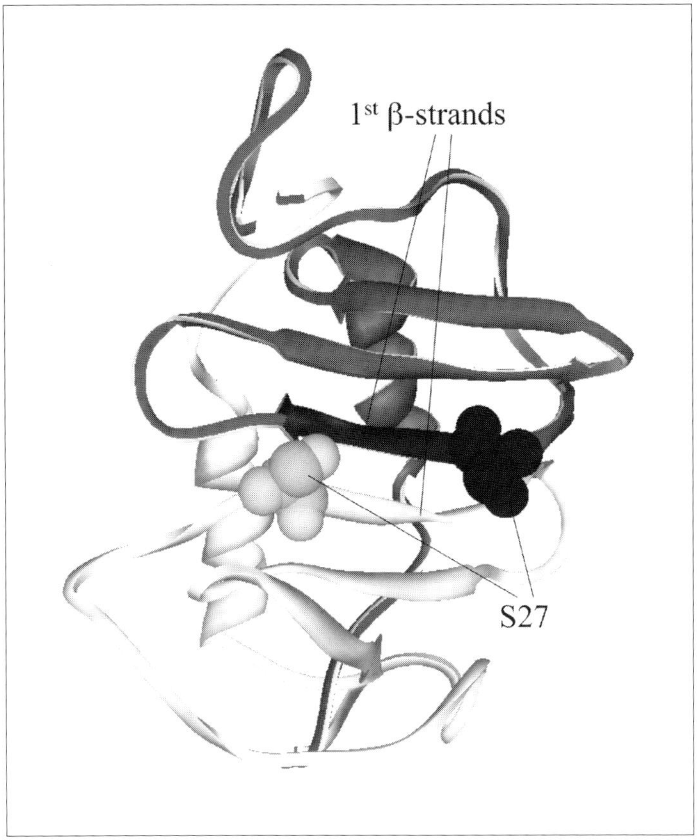

Figure 1
The homomeric dimer of CCL7 in a ribbon representation, using coordinates (1NCV) from the Protein Data Base [48] and Swiss Pdb Viewer [49] for display
The extended β-sheet composed by six β-strands (three from each subunit) faces the viewer. The two subunits are coloured light and medium grey, respectively. The first β-strands of both subunits, which form the primary dimer interface, are indicated. Residue S27 of CCL7 is indicated and shown with its van der Waals spheres in both subunits to illustrate the antiparallel orientation of the two first β-strands. S27 of CCL7 corresponds to E26 of CCL5, which is required for synergism between CCL5 and CXCL4 [44]. Note its orientation towards the opposite β-strand. In the subunit coloured dark grey, the first β-strand that mediates the synergistic activity of CCL7 together with CCL22 for CCR4 [46] is coloured dark grey, including residue S27. The CCL7 dimer was chosen for this representation for the reasons cited above and because it is formed through association of both subunits' β-sheets, similar to the homomeric complexes of CXCL4 and CXCL8 [42], and possibly their heteromeric complex [40, 41].

Conclusions

Based on the above, natural chemokine antagonism and synergism, as consequences of multiple and concomitant chemokine expression, constitute yet another level of regulation in leukocyte trafficking. By now, antagonism by protease-modified and native chemokines is well established as numerous cases have been documented in the last few years. A few recent reports have also illustrated its *in vivo* relevance, even though more studies on natural chemokine antagonists such as the pioneering work of Fulkerson et al. [31] would be desirable.

Overall, chemokine antagonism is clearly less frequent than chemokine agonism. While protease modification of chemokines and the resulting changes in chemokine activity have been investigated thoroughly, it remains for now unclear just how many more native antagonists exist. In fact, there might not be that many: we have screened roughly one third of more than 700 possible combinatorial chemokine-receptor combinations using chemotaxis assays with receptor transfectants. In these experiments, we identified only one additional partial agonist (CCL22 for CCR3) and one antagonist (CCL23 for CCR5) in addition to those already published (Petkovic V, Moghini C, and Gerber BO, unpublished observations). Hence, we think it unlikely that many more natural antagonists will be found. Rather, we would expect future breakthroughs in this area to stem from investigations into their physiologic or therapeutic relevance.

Compared to chemokine antagonism, the field of chemokine synergism is still in its infancy, even though the first report dates from more than a decade ago [43]. That chemokine and chemoattractant receptors can engage in cross talk with each other – or with members of other receptor classes – may not be too surprising, considering that this phenomenon has previously been reported for other G protein coupled receptors. It is likely that the recent reports mentioned above will trigger an increased interest for this topic in the chemokine community.

The occurrence of "cross-talk" between the (chemokine) ligands themselves is, in our opinion, the most exciting of all developments discussed here. What seems clear so far is that chemokine heteromers can be more potent agonists than the respective chemokines (or their homomers) alone, and that many but not all chemokines can induce or are susceptible to synergism. Even though we are at an early stage and have a limited understanding only of how these heteromers are formed, three lines of future research are evident already. For one, the chemokine as well as the receptor specificities of synergistic interactions will have to be assessed systematically and comprehensively, which will require diligence more than anything else. Second, the molecular and cellular reasons for the increased potency of synergistic heterodimers must be elucidated. Collectively, current evidence (see above and Fig. 1) implicates the first β-strand as an important mediator of synergism, furnishing a promising starting point for structure–function studies into chemokine synergism. Likewise, it will be important to determine if synergistic complexes induce receptor signalling or trafficking events that differ significantly from those elicited by the known agonist chemokines alone, and could thereby cause their increased activities. Last but not least, the *in vivo* relevance of chemokine synergism will need to be determined. These may seem daunting tasks, considering the multitude of chemokines, chemokine receptors, and target cells. Still, we trust that synergistic chemokines will continue to hold our attention and surprise us again in the future.

References

1 Struyf S, Proost P, Van Damme J (2003) Regulation of the immune response by the interaction of chemokines and proteases. *Adv Immunol* 81: 1–44

2 Proost P, Struyf S, Couvreur M, Lenaerts JP, Conings R, Menten P, Verhaert P, Wuyts A, Van Damme J (1998) Posttranslational modifications affect the activity of the human monocyte chemotactic proteins MCP-1 and MCP-2: identification of MCP-2(6-76) as a natural chemokine inhibitor. *J Immunol* 160: 4034–4041

3 Proost P, De Meester I, Schols D, Struyf S, Lambeir AM, Wuyts A, Opdenakker G, De Clercq E, Scharpe S, van Damme J (1998) Amino-terminal truncation of chemokines by CD26/dipeptidyl-peptidase IV. Conversion of RANTES into a potent inhibitor of monocyte chemotaxis and HIV-1-infection. *J Biol Chem* 273: 7222–7227

4 Struyf S, De Meester I, Scharpe S, Lenaerts JP, Menten P, Wang JM, Proost P, Van Damme J (1998) Natural truncation of RANTES abolishes signaling through the CC chemokine receptors CCR1 and CCR3, impairs its chemotactic potency and generates a CC chemokine inhibitor. *Eur J Immunol* 28: 1262–1271

5 Struyf S, Proost P, Schols D, De Clercq E, Opdenakker G, Lenaerts JP, Detheux M, Parmentier M, De Meester I, Scharpe S et al (1999) CD26/dipeptidyl-peptidase IV down-regulates the eosinophil chemotactic potency, but not the anti-HIV activity of human eotaxin by affecting its interaction with CC chemokine receptor 3. *J Immunol* 162: 4903–4909

6 Proost P, Struyf S, Schols D, Opdenakker G, Sozzani S, Allavena P, Mantovani A, Augustyns K, Bal G, Haemers A et al (1999) Truncation of macrophage-derived chemokine by CD26/ dipeptidyl-peptidase IV beyond its predicted cleavage site affects chemotactic activity and CC chemokine receptor 4 interaction. *J Biol Chem* 274: 3988–3993

7 McQuibban GA, Gong JH, Tam EM, McCulloch CA, Clark-Lewis I, Overall CM (2000) Inflammation dampened by gelatinase A cleavage of monocyte chemoattractant protein-3. *Science* 289: 1202–1206

8 Hensbergen PJ, van der Raaij-Helmer EM, Dijkman R, van der Schors RC, Werner-Felmayer G, Boorsma DM, Scheper RJ, Willemze R, Tensen CP (2001) Processing of natural and recombinant CXCR3-targeting chemokines and implications for biological activity. *Eur J Biochem* 268: 4992–4999

9 Proost P, Schutyser E, Menten P, Struyf S, Wuyts A, Opdenakker G, Detheux M, Par-

mentier M, Durinx C, Lambeir AM et al (2001) Amino-terminal truncation of CXCR3 agonists impairs receptor signaling and lymphocyte chemotaxis, while preserving antiangiogenic properties. *Blood* 98: 3554–3561

10 McQuibban GA, Gong JH, Wong JP, Wallace JL, Clark-Lewis I, Overall CM (2002) Matrix metalloproteinase processing of monocyte chemoattractant proteins generates CC chemokine receptor antagonists with anti-inflammatory properties *in vivo*. *Blood* 100: 1160–1167

11 Christopherson KW 2nd, Hangoc G, Broxmeyer HE (2002) Cell surface peptidase CD26/dipeptidylpeptidase IV regulates CXCL12/stromal cell-derived factor-1 alpha-mediated chemotaxis of human cord blood $CD34^+$ progenitor cells. *J Immunol* 169: 7000–7008

12 Ludwig A, Schiemann F, Mentlein R, Lindner B, Brandt E (2002) Dipeptidyl peptidase IV (CD26) on T cells cleaves the CXC chemokine CXCL11 (I-TAC) and abolishes the stimulating but not the desensitizing potential of the chemokine. *J Leukoc Biol* 72: 183–191

13 Moser B, Wolf M, Walz A, Loetscher P (2004) Chemokines: multiple levels of leukocyte migration control. *Trends Immunol* 25: 75–84

14 Clark-Lewis I, Kim KS, Rajarathnam K, Gong JH, Dewald B, Moser B, Baggiolini M, Sykes BD (1995) Structure-activity relationships of chemokines. *J Leukoc Biol* 57: 703–711

15 Loetscher P, Clark-Lewis I (2001) Agonistic and antagonistic activities of chemokines. *J Leukoc Biol* 69: 881–884

16 Loetscher P, Pellegrino A, Gong JH, Mattioli II, Loetscher M, Bardi G, Baggiolini M, Clark-Lewis II (2000) The ligands of CXC chemokine receptor 3, I-TAC, Mig and IP10, are natural antagonists for CCR3. *J Biol Chem* 10: 10

17 Blanpain C, Migeotte I, Lee B, Vakili J, Doranz BJ, Govaerts C, Vassart G, Doms RW, Parmentier M (1999) CCR5 binds multiple CC-chemokines: MCP-3 acts as a natural antagonist. *Blood* 94: 1899–1905

18 Nibbs RJ, Salcedo TW, Campbell JD, Yao XT, Li Y, Nardelli B, Olsen HS, Morris TS, Proudfoot AE, Patel VP et al (2000) C-C chemokine receptor 3 antagonism by the beta-chemokine macrophage inflammatory protein 4, a property strongly enhanced by an amino- terminal alanine-methionine swap. *J Immunol* 164: 1488–1497

19 Ogilvie P, Bardi G, Clark-Lewis I, Baggiolini M, Uguccioni M (2001) Eotaxin is a natural antagonist for CCR2 and an agonist for CCR5. *Blood* 97: 1920–1924

20 Martinelli R, Sabroe I, LaRosa G, Williams TJ, Pease JE (2001) The CC chemokine eotaxin (CCL11) is a partial agonist of CC chemokine receptor 2b. *J Biol Chem* 276: 42957–42964

21 Chou CC, Fine JS, Pugliese-Sivo C, Gonsiorek W, Davies L, Deno G, Petro M, Schwarz M, Zavodny PJ, Hipkin RW (2002) Pharmacological characterization of the chemokine receptor, hCCR1 in a stable transfectant and differentiated HL-60 cells: antagonism of hCCR1 activation by MIP-1beta. *Br J Pharmacol* 137: 663–675

22 Wan Y, Jakway JP, Qiu H, Shah H, Garlisil CG, Tian F, Ting P, Hesk D, Egan RW, Billah MM et al (2002) Identification of full, partial and inverse CC chemokine receptor 3 agonists using [35S]GTPgammaS binding. *Eur J Pharmacol* 456: 1–10

23 Xanthou G, Duchesnes CE, Williams TJ, Pease JE (2003) CCR3 functional responses are regulated by both CXCR3 and its ligands CXCL9, CXCL10 and CXCL11. *Eur J Immunol* 33: 2241–2250
24 Ogilvie P, Paoletti S, Clark-Lewis I, Uguccioni M (2003) Eotaxin-3 is a natural antagonist for CCR2 and exerts a repulsive effect on human monocytes. *Blood* 102: 789–794
25 Petkovic V, Moghini C, Paoletti S, Uguccioni M, Gerber B (2004) I-TAC/CXCL11 is a natural antagonist for CCR5. *J Leukoc Biol* 76: 701–708
26 Petkovic V, Moghini C, Paoletti S, Uguccioni M, Gerber B (2004) Eotaxin-3/CCL26 is a natural antagonist for CC chemokine receptors 1 and 5. A human chemokine with a regulatory role. *J Biol Chem* 279: 23357–23363
27 Parody TR, Stone MJ (2004) High level expression, activation, and antagonism of CC chemokine receptors CCR2 and CCR3 in Chinese hamster ovary cells. *Cytokine* 27: 38–46
28 Ogilvie P, Thelen S, Moepps B, Gierschik P, da Silva Campos AC, Baggiolini M, Thelen M (2004) Unusual chemokine receptor antagonism involving a mitogen-activated protein kinase pathway. *J Immunol* 172: 6715–6722
29 Gong JH, Ratkay LG, Waterfield JD, Clark-Lewis I (1997) An antagonist of monocyte chemoattractant protein 1 (MCP-1) inhibits arthritis in the MRL-lpr mouse model. *J Exp Med* 186: 131–137
30 Sasaki M, Hasegawa H, Kohno M, Inoue A, Ito MR, Fujita S (2003) Antagonist of secondary lymphoid-tissue chemokine (CCR ligand 21) prevents the development of chronic graft-versus-host disease in mice. *J Immunol* 170: 588–596
31 Fulkerson PC, Zimmermann N, Brandt EB, Muntel EE, Doepker MP, Kavanaugh JL, Mishra A, Witte DP, Zhang H, Farber JM et al (2004) Negative regulation of eosinophil recruitment to the lung by the chemokine monokine induced by IFN-gamma (Mig, CXCL9). *Proc Natl Acad Sci USA* 101: 1987–1992
32 Struyf S, Proost P, Lenaerts JP, Stoops G, Wuyts A, Van Damme J (2001) Identification of a blood-derived chemoattractant for neutrophils and lymphocytes as a novel CC chemokine, Regakine-1. *Blood* 97: 2197–2204
33 Struyf S, Stoops G, Van Coillie E, Gouwy M, Schutyser E, Lenaerts JP, Fiten P, van Aelst I, Proost P, Opdenakker G et al (2001) Gene cloning of a new plasma CC chemokine, activating and attracting myeloid cells in synergy with other chemoattractants. *Biochemistry* 40: 11715–11722
34 Gouwy M, Struyf S, Mahieu F, Put W, Proost P, Van Damme J (2002) The unique property of the CC chemokine regakine-1 to synergize with other plasma-derived inflammatory mediators in neutrophil chemotaxis does not reside in its NH2-terminal structure. *Mol Pharmacol* 62: 173–180
35 Gouwy M, Struyf S, Catusse J, Proost P, Van Damme J (2004) Synergy between proinflammatory ligands of G protein-coupled receptors in neutrophil activation and migration. *J Leukoc Biol* 76: 185–194
36 Reca R, Mastellos D, Majka M, Marquez L, Ratajczak J, Franchini S, Glodek A, Honczarenko M, Spruce LA, Janowska-Wieczorek A et al (2003) Functional receptor for C3a

anaphylatoxin is expressed by normal hematopoietic stem/progenitor cells, and C3a enhances their homing-related responses to SDF-1. Blood 101: 3784–3793

37 Krug A, Uppaluri R, Facchetti F, Dorner BG, Sheehan KC, Schreiber RD, Cella M, Colonna M (2002) IFN-producing cells respond to CXCR3 ligands in the presence of CXCL12 and secrete inflammatory chemokines upon activation. *J Immunol* 169: 6079–6083

38 Vanbervliet B, Bendriss-Vermare N, Massacrier C, Homey B, de Bouteiller O, Briere F, Trinchieri G, Caux C (2003) The inducible CXCR3 ligands control plasmacytoid dendritic cell responsiveness to the constitutive chemokine stromal cell-derived factor 1 (SDF-1)/CXCL12. *J Exp Med* 198: 823–830

39 Kohrgruber N, Groger M, Meraner P, Kriehuber E, Petzelbauer P, Brandt S, Stingl G, Rot A, Maurer D (2004) Plasmacytoid dendritic cell recruitment by immobilized CXCR3 ligands. *J Immunol* 173: 6592–6602

40 Dudek AZ, Nesmelova I, Mayo K, Verfaillie CM, Pitchford S, Slungaard A (2003) Platelet factor 4 promotes adhesion of hematopoietic progenitor cells and binds IL-8: novel mechanisms for modulation of hematopoiesis. *Blood* 101: 4687–4694

41 Nesmelova IV, Sham Y, Dudek AZ, van Eijk LI, Wu G, Slungaard A, Mortari F, Griffioen AW, Mayo KH (2004) Platelet factor 4 and interleukin-8 CXC-chemokine heterodimer formation modulates function at the quarternary structural level. *J Biol Chem* 280: 4948–4958

42 Lortat-Jacob H, Grosdidier A, Imberty A (2002) Structural diversity of heparan sulfate binding domains in chemokines. *Proc Natl Acad Sci USA* 99: 1229–1234

43 Broxmeyer HE, Sherry B, Cooper S, Lu L, Maze R, Beckmann MP, Cerami A, Ralph P (1993) Comparative analysis of the human macrophage inflammatory protein family of cytokines (chemokines) on proliferation of human myeloid progenitor cells. Interacting effects involving suppression, synergistic suppression, and blocking of suppression. *J Immunol* 150: 3448–3458

44 Von Hundelshausen P, Koenen RR, Sack M, Mause SF, Adriaens W, Proudfoot AE, Hackeng TM, Weber C (2004) Heterophilic interactions of platelet factor 4 and RANTES promote monocyte arrest on endothelium. *Blood* 105: 924–930

45 Paoletti S, Petkovic V, Sebastiani S, Danelon MG, Uguccioni M, Gerber BO (2005) A rich chemokine environment strongly enhances leukocyte migration and activities. *Blood* 105: 3405–3412

46 Sebastiani S, Danelon MG, Gerber BO, Uguccioni M (2005) CCL22-induced responses are powerfully enhanced by synergy inducing chemokines via CCR4: evidence for the involvement of the first beta-strand of chemokine. *Eur J Immunol* 35: 746–756

47 Rodriguez-Frade JM, Mellado M, Martinez AC (2001) Chemokine receptor dimerization: two are better than one. *Trends Immunol* 22: 612–617

48 Berman HM, Westbrook J, Feng Z, Gilliland G, Bhat TN, Weissig H, Shindyalov IN, Bourne PE (2000) The Protein Data Bank. *Nucleic Acids Res* 28: 235–242

49 Guex N, Peitsch MC (1997) SWISS-MODEL and the Swiss-PdbViewer: an environment for comparative protein modeling. *Electrophoresis* 18: 2714–2723

Effector cell traffic-unrelated functions

Crosstalk between chemokine, opioid, and vanilloid receptors

Ning Zhang[1,2] *and Joost J. Oppenheim*[1]

[1]Laboratory of Molecular Immunoregulation, Intramural Research Support Program, Building 560, Room 21-89A, Frederick, MD 21702-1201, USA; [2]Department of Chemical Biology, and State Key Laboratory of Molecular Dynamic and Stable Structures, College of Chemistry, Peking University, Beijing 100871, China

Introduction

Increasing evidence indicates that the immune and neural systems interact by a wide variety of mechanisms [1]. The tight blood–brain barrier (BBB) restricts the communication between central nervous system (CNS) and immune system, and protects the brain from the damaging effects of inflammation. Nevertheless, multiple interconnections exist between these two systems. (1) The autonomic nervous system is embedded in many peripheral sites along with the immune system, such as the liver, spleen, bone marrow, thymus, lymph nodes, skin, and gastrointestinal tract. (2) Neurotransmitters produced by stimulation of the sympathetic and parasympathetic system directly influence leukocyte function. For example, acetylcholine, norepinephrine, and Met-encephalin suppress cells engaged in both innate and adaptive immunity [2–4]. In contrast, calcitonin gene-related peptide (CGRP) and substance P released by pain fibers enhance inflammation [5, 6]. (3) The CNS can also suppress immune response by activation of the hippocampal-pituitary-adrenal (HPA) axis. In response to environmental stress, corticotropin-releasing factor (CRF) secreted by the hippocampus activates the pituitary to produce adrenocorticotropin hormones (ACTH) [7]. ACTH in turn activates the adrenals to produce corticosterones; hormones with potent immune suppressive effects. (4) There is also evidence for the existence of highly localized "windows" in the blood-brain barrier, called circumventricular organs. These "windows" allow transmission of soluble mediators released by immune cells to enter the hypothalamus of the brain [8].

We have studied the role of receptor cross-talk in the communication between immune and neural systems. Receptors that are essential for immune system functions have been detected on neuronal cells, and typically neuronal receptors are also

Chemokine Biology – Basic Research and Clinical Application, Volume I
edited by Bernhard Moser, Gordon L. Letts and Kuldeep Neote

expressed by peripheral leukocytes. Activation of one receptor often causes an alternation in the function of nearby other receptors expressed on the same cells. For example, opioid receptors, the key neuronal analgesic receptors, have also been detected on leukocytes. Prolonged activation of opioid receptors on leukocytes dampens chemokine receptor responses [4]. In contrast, chemokine receptors are expressed on peripheral sensory neurons and in the CNS. As will be discussed, chemokines are capable of reversing opioid receptor-mediated analgesic effects [9, 10]. The receptors for prostaglandins and bradykinins, two proinflammatory mediators, are also expressed on sensory neurons. Activation of these receptors enhances the perception of pain by increasing the sensitivity of the Vanilloid receptor 1 (TRPV1), a pain receptor, expressed on the same sensory neurons [11–13]. The Vanilloid receptors in the oral cavity have the capacity to respond to capsaicin, spicy components of peppers. In addition, proinflammatory chemokines are also capable of sensitizing TRPV1 by phosphorylation of its Ser/Thr residues. Conversely, it is clearly documented that secretion of CGRP and Substance P from Vanilloid receptor-activated sensory neurons has proinflammatory effects [5, 6]. Further characterization is underway to map the expression of TRPV receptor family members on immune cells and to determine if they have a role in regulating chemokine receptors. In this chapter, we will focus on the bi-directional desensitization between chemokine and opioid receptors that reduces the perception of pain, and sensitization of Vanilloid receptor 1 (TRPV1) by chemokine receptors that promotes pain signals.

Opiates suppress immune responses

Opiates have long been used to suppress "pain" and enhance "pleasure" in human history. However, abusive usage of opiates leads to a greater prevalence of viral hepatitis, HIV infection, bacterial pneumonias, tuberculosis, CNS infection, and endocarditis [14–16]. These pathological conditions can be explained by opioid-induced suppression of a spectrum of immune host defenses. Chronic morphine administration induces lymphoid organ atrophy, loss of natural killer (NK) cell activity, and a diminished ratio of $CD4^+CD8^+$ cells in the thymus [17]. In rats, repetitive morphine treatment impairs the delayed hypersensitivity skin response to tuberculin [18]. Morphine also inhibits transcription of interferon γ in activated T cells, which may contribute to an increase in HIV infection among morphine users [19]. Chemokine receptor-mediated migration of human leukocytes was also compromised by *in vitro* pre-incubation of cells with opioids [20, 21]. In addition to these immunosuppressive effects, it has been reported that opioids also exhibit certain positive effects on immune responses, including enhanced synthesis of tumor necrosis factor-α and interleukin-1β by activated macrophages, and direct induction of leukocyte chemotaxis [4, 22].

Opioids induce immunosuppressive effects by enhancing neurohormone production

Prolonged activation of CNS by morphine leads to a 3- to 4-fold increases in the level of circulating corticosterone, up to 400–450 ng/ml, resulting in splenic and thymic atrophy, a decrease in lymphocyte proliferation, inhibition of IL-2 and IFN-γ synthesis [23]. Conversely, disruption of μ-opioid receptors blocks morphine induced increase in circulating corticosterone. The immunomodulatory effects of chronic morphine treatment are significantly attenuated in *mor*$^{-/-}$ mice. Supplemental infusion of corticosterone partially reproduces the immunodeficiency [24]. Opioids also activate the sympathetic nervous system, resulting in an increase in the level of circulating epinephrine from the adrenal medulla and norepinephrine from sympathetic nerve terminals [25]. Increased catecholamine levels have been linked to suppression of NK cell and lymphocyte function [26].

Opioids downregulate chemokine receptors by heterologous desensitization

Receptor desensitization is a key mechanism for protecting cells from prolonged responses to the agonists. The desensitization process of a GPCR can be initiated with its own ligand, causing homologous desensitization, or by activation of other "nearby" receptors, resulting in heterologous desensitization. Homologous desensitization mainly involves the activation of the feedback inhibitors, GRK and arrestins [27]. Heterologous desensitization is usually mediated by second messenger-activated kinases, such as PKA and PKC [28]. When the cytosolic tail of a GPCR is phosphorylated, the receptor loses its effective coupling to downstream G proteins, and sometimes even undergoes internalization, resulting in the loss of receptor function.

As discussed in previous chapters, chemokine receptors play a critical role in cell trafficking, development, activation of inflammatory and immune cells, and HIV infection. Upon injury, exogeneous microbial products, such as fMLP, and production of endogenous chemokines create an *in vivo* concentration gradient. Chemokine receptors on leukocytes sense the chemical gradient and direct the cells towards the inflammatory site. Chemokine receptors are coupled to Gi/o proteins. Consequently, PI3 kinases are recruited to the leading edge of a cell, which elicits a chain of downstream signaling events, including activation of CDC42/Rac, recruitment of Arp2/3 complex, and assembly of actin filaments. Formation of the actin filaments in the front of a cell is believed to be the driving force of chemotaxis [29]. Chemokine receptors also mediate other signaling pathways, such as G-protein dependent activation of phospholipase C and protein kinase C, and G-protein independent recruitment of G protein coupled receptor kinases (GRK) and arrestins. All three subtypes of opioid receptors, identical to their counterparts in the brain, are co-expressed by leukocytes along with chemokine receptors [30]. Although opioids

Table 1 - Crosstalk between chemokine, opioid, and vanilloid receptors

Effecter receptors	Cell types	Target receptors	Effects
MOR, DOR	Leukocytes	Desensitize CCR1, CCR2, CCR5, CXCR1/2	Immuno-suppression [4, 20–22, 30]
CCR1, CCR2, CCR7, CXCR4, CXCR1/2, CCR5	Neurons and leukocytes	Desensitize MOR, DOR	Hyperalgesia [4, 9, 10]
CCR1, CCR2, CCR5, CXCR1/2	Neurons	Sensitize TRPV1	Hyperalgesia

Abbreviations: MOR, μ-opioid receptors; DOR, δ-opioid receptors. TRPV1, Transient receptor potential vanilloid 1, also called vanilloid receptor

exhibit a moderate capacity to induce opioid receptor-dependent chemotaxis *in vitro*, their principal effect is to suppress inflammation by inhibiting chemokine receptor function [20–21]. Pretreatment with opioids selectively inhibits a number of chemokine receptors, including CCR1, CCR2, CXCR1 and CXCR2 on myeloid cells, such as human monocytes and neutrophils (Tab. 1). Additional studies reveal similar opioid-induced heterologous desensitization of chemokine receptors on T-lymphocytes.

Heterologous desensitization of chemokine receptors involves uncoupling of Gi protein by calcium-independent PKC

Met-enkephalin stimulation of opioid receptors activates phospholipase Cβ, resulting in the accumulation of IP3 and diacylglycerol (DAG) from PIP2 (4,5) hydrolysis (Fig. 1) [31]. This opioid-induced production of IP3 is rather modest, as indicated by the lack of transient calcium influx. At the same time, the capacity of Met-enkephalin to induce chemotaxis suggests that PI3 kinase γ is activated as well. Both DAG and PI3 kinase γ activate Protein Kinase C, a family of Ser/Thr kinases. The 12 PKC isozymes can be divided into three subfamilies based on differences in activation: classical PKCs (cPKCs), such as α, βI, βII, and γ, require both Ca^{2+} and DAG for activation; novel PKCs (nPKCs), such as δ, ε, θ, and η, are DAG-dependent but Ca^{2+}-independent; and atypical PKCs, such as ζ and λ, require neither Ca^{2+} nor DAG [32]. Recent studies have suggested that atypical PKCs may be activated by PI3 kinases [33]. Eight PKC isozymes, α, β1, β2, δ, ε, η, μ, and ζ, have been identified in human blood monocytes. Biochemical analysis of human monocytes and HEK cells transfected to express μ-opioid receptors (MOR) and chemokine receptors reveals that opioid induced heterologous desensitization involved calcium-inde-

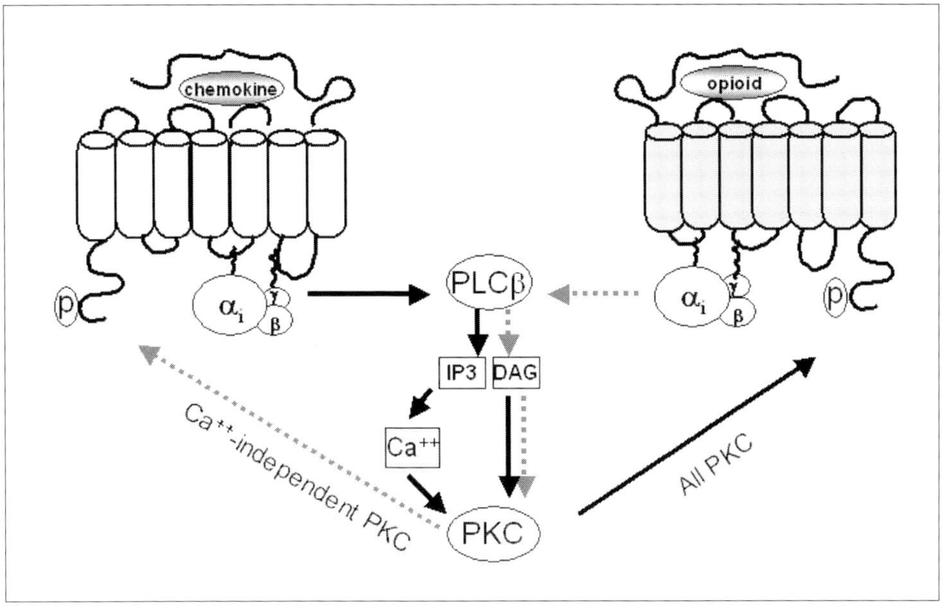

Figure 1
Molecular mechanism of bi-directional heterologous desensitization between chemokine and opioid receptors. In leukocytes, opioids induce heterologous desensitization of chemokine receptors through Gi proteins, phospholipase Cβ (PLCβ), and Ca^{2+}-independent protein kinase C (PKC), resulting in an immunosuppressive effect. In sensory neurons, treatment with proinflammatory chemokines downregulates opioid receptor function through both Ca^{2+}-dependent and -independent protein kinase C, resulting in hyperalgesia. Phosphorylation of the cytoplasmic tail and intracellular loops of a seven-transmembrane receptor by PKC decouples the receptor from downstream Gi-proteins, resulting in a decrease in receptor function. (IP3, inositol 1, 4, 5-triphosphate; DAG, diacylglycerol)

pendent PKC [21]. Activation of PKC is associated with the enhanced phosphorylation of chemokine receptors, resulting in a decrease in their affinity and in reduced coupling to G-proteins. Consequently, chemokine receptor mediated chemotaxis, calcium influx, and HIV infection are impaired.

Opioid-induced heterologous desensitization exhibits selectivity

In human monocytes, only μ and δ opioid receptors were detected to inhibit chemokine receptors [20]. Furthermore, opioid treatment inhibits the chemotactic response of human monocytes and neutrophil to a limited selection of chemokines, including IL-8, MIP-1α, RANTES, and MCP-1, but not NAP-1, MIP-1β, SDF-1α, or fMLP [4, 22]. The availability of chemokine receptors to be desensitized may be based on their intrinsic properties: the accessibility of their C-terminal tails to phosphorylation, the impact of phosphorylation on their capacity to activate G-protein, and/or the activation threshold of each chemokine receptor. Chemokine receptors are arranged in a hierarchy in their capacity to induce heterologous desensitization [28]. For example, certain receptors, such as the fMLP receptor, have a higher capacity to desensitize other GPCRs than to be desensitized. Treatment with fMLP causes a greater phosphorylation and internalization of C5a and IL8 receptors, resulting in over 50% inhibition of their function. In contrast, IL8 has lower inhibitory effects on fMLP receptors. The capacity of a receptor to cross-desensitize GPCRs seems to correlate with its ability to induce greater phosphoinositide hydrolysis and sustained calcium mobilization [21, 28]. Opioid induced heterologous desensitization has only modest inhibitory effects on leukocyte chemotactic responses. The lower inhibitory effects are probably due to a lower expression of opioid receptors on leucocytes than on certain neuronal cells, resulting in a limited activation of downstream PKC [21].

Chemokines inhibit opioid receptors on leukocytes and sensory neurons

Pretreatment of monocytes with chemokines inhibits δ- and μ-opioid receptor mediated chemotaxis [10]. The inhibitory effects are elicited by ligand activation of selective chemokine receptors, including CCR2, CCR5, CCR7, and CXCR4, but not by CXCR1 or CXCR2. The heterologous desensitization of opioid receptors by chemokine receptors is also mediated by Gi protein mediated protein kinase C activation. Prolonged treatment with chemokines induces the phosphorylation of MOR, resulting in loss of surface MOR via receptor internalization and uncoupling of MOR from downstream effector G proteins [9]. The pathophysiological relevance of chemokine-induced desensitization of opioid receptors on leukocytes is unclear. We therefore decided to consider whether chemokine receptors expressed on neuronal cells desensitized nearby opioid receptors.

Expression of chemokine receptors in the central and peripheral nervous system

Many chemokine receptors, with the exception of CCR6 and 7, have been reported to be normally expressed by cells of the CNS, including astrocytes, microglial cells, oligodendrocytes, and neurons [34]. Chemokines and their receptors in the CNS participate in pathological, inflammatory, and neurodegenerative conditions, such as multiple sclerosis, experimental autoimmune encephalitis, Alzheimer's disease,

HIV infection, demential complex, brain injury, and tumors. Furthermore, chemokines are also involved in brain development [35]. Knockout of the mouse gene for CXCR4 or its ligand CXCL12 causes the disruption of the laminar architecture, probably due to a premature and disorganized inward migration of external granular layer cells [36]. Chemokines also indirectly regulate neuronal signaling. For example, high levels of KC, the murine homolog of CXCL1, cause a progressive neurological dysfunction, characterized by ataxia, postural instability, and rigidity [37]. Furthermore, CXCL8 and CXCL12 enhance synaptic activity by increasing neurotransmitter release and suppressing the induction of long-term depression [38]. On the other hand, soluble CX3CL1/fractalkine was able to reduce calcium oscillations in synoptically coupled hippocampal neurons by decreasing glutamate secretion and blocked gp120-induced apoptosis [39]. Thus, there is considerable evidence for the expression of various functional chemokine receptors by neuronal cells.

Molecular mechanism of opioid receptors-mediated analgesic effects

Opioid receptors consist of a family of seven-transmembrane receptors, with three subtypes, μ, δ, and κ [31]. They exert analgesic effects by blocking either the sensing or the propagation of pain signals. Endogenous peptides, such as endorphins and Met-enkephalin, have been shown to bind to opioid receptors and to exert an analgesic effect similar to that of morphine, heroin, and other plant extracts, indicating that opioids and their receptors provide an intrinsic mechanism to enable a host to perceive "pain" and "pleasure". Binding of opioids induces a conformational change in the receptors and causes the dissociation of heterotrimeric Gi/o proteins immediately downstream of the opioid receptors. Consequently, both Gα and Gβγ orchestrate a spectrum of downstream responses, including activation of G-protein coupled inward rectify potassium channel (GIRK), inhibition of adenylyl cyclase and various calcium channels. Activation of GIRK hyperpolarizes neuronal membranes, thereby preventing the excitation and transmission of pain signals. Inhibition of calcium channels impairs the release of neurotransmitters, which is also critical for the perception of pain. Furthermore, opioids also induce a transient calcium influx in both primary neurons and opioid-receptor-transfected cell lines, probably due to the activation of phospholipase C.

Pro-inflammatory chemokines suppress the function of opioid receptors

Chemokine receptors are detected on the same neuronal cells expressing opioid receptors [9]. Immunohistochemical staining shows the co-expression of CCR1 and MOR on sensory neurons in rat dorsal root ganglion. Several proinflammatory chemokines, such as CCL2, CCL3, CCL5 and CXCL8, are able to induce a transient but robust calcium influx in a subpopulation of sensory neurons, indicating that these neuronal chemokine receptors are functional [40]. Pretreatment of sensory neurons from rat dorsal root ganglion by these chemokines downregulates the function of MOR. The molecular mechanism of chemokine-induced heterologous desensitization of MOR was further investigated in a HEK293 cell line transfected to stably express both MOR and CCR1. CCL3 treatment causes marked inhibitory effects by phosphorylating the receptors, decoupling MOR from G protein, followed by internalization of MOR. Thus, chemokine induced heterologous desensitization of MOR on sensory neurons is also dependent on Gi-mediated activation of PKC (Fig. 1). The *in vitro* observation on desensitization of MOR on primary sensory neurons was confirmed by a cold-water tail flick assay [10]. Introduction of a specific ligand for MOR, DAMGO, into the rat periaqueductal gray center (PAG) significantly enhances the tail-flick latency, indicative of MOR-mediated analgesic effects. Pre-administration of chemokines impaired the DAMGO-induced analgesic effects, suggesting chemokine-induced heterologous desensitization of opioid receptor restores the sensing of pain [9, 10].

Cross-talk between "pain" and chemokine receptors

Since the crosstalk between chemokine and opioid receptors resulted in increased pain perception, we wondered whether there also would be any crosstalk between chemokine and pro-pain receptors. Painful signals are detected by a group of specialized sensory neurons called nociceptors [41]. Recently, the first "pain" receptor, TRPV1 (vanilloid receptor 1, VR1), was identified to be a ligand-gated six-transmembrane calcium channel, highly expressed in nociceptors [42]. Noxious stimuli, such as capsaicin, heat, cold, pressure, acid, and inflammatory mediators, induce the opening of this calcium channel. As a consequence, the membrane is depolarized and the action potential is propagated to the CNS as a pain signal. It has been well documented since ancient Greece that inflammation enhances pain and that pain represents another host defense mechanism. A variety of cellular mediators, such as bradykinin, nerve growth factor, and prostaglandins (PGE2), have been shown to contribute to hyperalgesia by regulating the expression, sensitivity, and desensitization of TRPV1 [41]. Bradykinin, a potent inflammatory mediator, does so by inducing the production of endogenous "pain" ligand, 12-HPETE [43]. Inflammation elicits the accumulation of nerve growth factor (NGF) and activation of p38 MAPK, resulting in the enhancement of the translation of TRPV1 in primary neurons [44]. Nerve growth factor (NGF), a member of the interleukin 1 family, can also sensitize TRPV1 by inducing hydrolysis of PtdIns(4,5)P2, an inhibitor of TRPV1 [45]. PGE2, by coupling to Gs, induces the phosphorylation of TRPV1 by PKA, resulting in a significant decrease in desensitization, i.e., TRPV1 maintains sensitivity despite repetitive stimulation [46]. Thus TRPV1 is an appropriate target for chemokine receptor signals.

Chemokine receptors sensitize TRPV1 on sensory neurons

The expression pattern of CCR1 partially overlaps that of TRPV1 on the sensory neurons of dorsal root ganglion and about 39±3% of DRG neurons express both receptors. Chemokine receptors have been proposed to directly contribute to the inflammation-induced hyperalgesia by inducing a transient calcium influx in neuronal cells [40]. Such a chemokine-induced calcium influx is capable of eliciting an action potential. However, we consider it unlikely that any neuronal calcium influx will result in the perception of pain, since opioids which also induce neuronal calcium influx are far from painful [9]. Pretreatment with CCL3 enhanced the sensitivity of TRPV1 to capsaicin by three- to five-fold as measured by calcium flux responses *in vitro*. The sensitization effects are likely due to the removal of PIP2, a TRPV1 endogenous inhibitor, and phosphorylation of the calcium channel by PKC (Fig. 2). Intrathecal injection of CCL3 to the spinal cord enhanced the rate of mouse hind paw withdrawal from the painful stimulation by heat, indicating the relevance of the *in vitro* observation. The fact that a proinflammatory chemokine, by interacting with its receptor on small diameter neurons, indirectly sensitizes TRPV1 suggests that the process of receptor cross-sensitization may contribute to hyperalgesia during inflammation.

Effects of activation of TRPV receptors on inflammatory responses

Opening of TRPV1 calcium channel induces the production and secretion of calcitonin gene-related peptide (CGRP) and Substance P, two potent neuropeptides regulating leukocyte function [5, 6]. CGRP in the airways causes vasodilatation, and in a few instances, bronchoconstriction. It also induces eosinophil migration, stimulates secretion of cytokines from antigen-specific T cells, and enhances of beta-integrin-mediated T cell adhesion to fibronectin at the site of inflammation. On the other hand, CGRP also impairs the capacity of macrophages to activate T-cells, a potent anti-inflammatory effect. Substance P acts through NK1 receptor expressed on T cells, macrophages, dendritic cells and probably other cell types, resulting in an increase in IFN-γ production and amplification of the Th1 response. TRPV1 may also directly modulate leukocyte function. Treatment of T cells with capsaicin inhibits IkappaB kinase activation, resulting in impaired T cell activation [47]. Whether the cross-talk between chemokine and TRPV receptors on leukocytes is bi-directional remains to be determined. Although painful stimuli may promote inflammatory host defenses, the net effects of TRPV receptor on the immune system

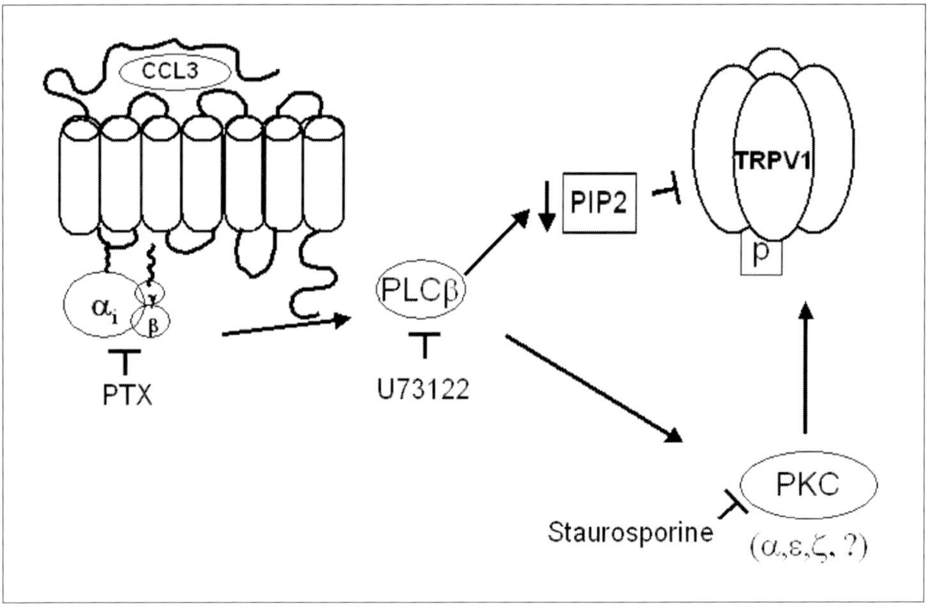

Figure 2
Molecular mechanism of chemokine-induced sensitization of Vanilloid receptor 1 (TRPV1). Activation of CCR1 by CCL3/MIP-1a enhances the sensitivity of TRPV1, a "pain" receptor, through a signal transduction cascade involving Gi protein, phospholipase Cβ (PLCβ), and protein kinase C (PKC). PLCβ hydrolyzes phosphoinositol 4,5-biphosphate (PIP2), an endogenous inhibitor of TRPV1, thereby sensitizing the TRPV1 pain receptor. Phosphorylation of TRPV1 by PKC enhances the sensitivity of TRPV1. CCL3/MIP-1α-induced sensitization of TRPV1 can be blocked at various steps of the signaling cascade by pertussis toxin (PTX), U73122, or staurosporine.

are still not clear. Further *in vivo* and *in vitro* investigations are needed to establish the pathophysiological relevance of the cross-talk between TRPV and chemokine receptors.

Conclusions

Cross-talk between chemokine and neuronal receptors provides a mechanism for integrating neuronal and immune responses. Chemokine receptors play a pivotal role during this communication. Pretreatment with opioids induces heterologous desensitization of chemokine receptors on leukocytes by activating Gi proteins and calcium-independent PKC. Conversely, chemokines also desensitize neuronal receptors for opioids, which enhance pain perception. Furthermore, exposure to chemokines sensitizes TRPV1 "pain" receptors which generate a "painful" signal from sensory neurons to the host CNS. Both of the opioid and Vanilloid pathways warn the host of the existence of a pathological condition. In the future, it will be interesting to investigate the communication between chemokines and neuronal responses in several disease settings. For example, herpes zoster and rheumatoid arthritis are extremely painful inflammatory diseases. Blocking chemokine receptors may significantly reduce the painful symptom. Furthermore, a decrease in nociceptive neuron activity will in turn reduce the secretion of proinflammatory neurotransmitters, such as CGRP and Substance P. Therefore, blocking proinflammatory chemokines may serve as an effective approach to block the positive feedback loops between inflammation and hyperalgesia.

Acknowledgement

We thank Dr. Thomas J. Rogers and Dr. O.M. Zack Howard for inspiring discussion.

References

1 Steinman L (2004) Elaborate interactions between the immune and nervous systems. *Nat Immunol* 5(6): 575–581

2 Tracey KJ (2002) The inflammatory reflex. *Nature* 420(6917): 853–859

3 Borovikova LV, Ivanova S, Zhang M, Yang H, Botchkina GI, Watkins LR, Wang H, Abumrad N, Eaton JW, Tracey KJ (2000) Vagus nerve stimulation attenuates the systemic inflammatory response to endotoxin. *Nature* 405(6785):458–462

4 Rogers TJ, Steele AD, Howard OM, Oppenheim JJ (2000) Bidirectional heterologous desensitization of opioid and chemokine receptors. *Ann NY Acad Sci* 917:19–28

5 Springer J, Geppetti P, Fischer A, Groneberg DA (2003) Calcitonin gene-related peptide as inflammatory mediator. *Pulm Pharmacol Ther* 16(3):121–130

6 Weinstock JV (2004) The role of substance P, hemokinin and their receptor in governing mucosal inflammation and granulomatous responses. *Front Biosci* 9: 1936–1943

7 Webster JI, Sternberg EM (2004) Role of the hypothalamic-pituitary-adrenal axis, glucocorticoids and glucocorticoid receptors in toxic sequelae of exposure to bacterial and viral products. *J Endocrinol* 181(2): 207–221

8 Roth J, Harre EM, Rummel C, Gerstberger R, Hubschle T (2004) Signaling the brain in systemic inflammation: role of sensory circumventricular organs. *Front Biosci* 9: 290–300

9 Zhang N, Rogers TJ, Caterina M, Oppenheim JJ (2004) Proinflammatory chemokines, such as C-C chemokine ligand 3, desensitize mu-opioid receptors on dorsal root ganglia neurons. *J Immunol* 173(1): 594–599

10 Szabo I, Chen XH, Xin L, Adler MW, Howard OM, Oppenheim JJ, Rogers TJ (2002) Heterologous desensitization of opioid receptors by chemokines inhibits chemotaxis and enhances the perception of pain. *Proc Natl Acad Sci USA* 99(16): 10276–10281
11 Caterina MJ, Julius D (2001) The vanilloid receptor: a molecular gateway to the pain pathway. *Annu Rev Neurosci* 24: 487–517
12 Hu HJ, Bhave G, Gereau RW 4th (2002) Prostaglandin and protein kinase A-dependent modulation of vanilloid receptor function by metabotropic glutamate receptor 5: potential mechanism for thermal hyperalgesia. *J Neurosci* 22(17): 7444–7452
13 Chuang HH, Prescott ED, Kong H, Shields S, Jordt SE, Basbaum AI, Chao MV, Julius D (2001) Bradykinin and nerve growth factor release the capsaicin receptor from PtdIns(4,5)P2-mediated inhibition. *Nature* 411(6840): 957–962
14 Louria DB, Hensle T, Rose J (1967) The major medical complications of heroin addiction. *Ann Intern Med* 67(1): 1–22
15 Reichman LB, Felton CP, Edsall JR (1979) Drug dependence, a possible new risk factor for tuberculosis disease. *Arch Intern Med* 139(3): 337–339
16 Haverkos HW, Lange WR (1990) Serious infections other than human immunodeficiency virus among intravenous drug abusers. From the Alcohol, Drug Abuse, and Mental Health Administration. *J Infect Dis* 161(5): 894–902
17 Novick DM, Ochshorn M, Ghali V, Croxson TS, Mercer WD, Chiorazzi N, Kreek MJ (1989) Natural killer cell activity and lymphocyte subsets in parenteral heroin abusers and long-term methadone maintenance patients. *J Pharmacol Exp Ther* 250(2): 606–610
18 Pellis NR, Harper C, Dafny N (1986) Suppression of the induction of delayed hypersensitivity in rats by repetitive morphine treatments. *Exp Neurol* 93(1): 92–97
19 Wang J, Barke RA, Charboneau R, Loh HH, Roy S (2003) Morphine negatively regulates interferon-gamma promoter activity in activated murine T cells through two distinct cyclic AMP-dependent pathways. *J Biol Chem* 278(39): 37622–37631
20 Grimm MC, Ben-Baruch A, Taub DD, Howard OM, Resau JH, Wang JM, Ali H, Richardson R, Snyderman R, Oppenheim JJ (1998) Opiates transdeactivate chemokine receptors: delta and mu opiate receptor-mediated heterologous desensitization. *J Exp Med* 188(2): 317–325
21 Zhang N, Hodge D, Rogers TJ, Oppenheim JJ (2003) Ca^{2+}-independent protein kinase Cs mediate heterologous desensitization of leukocyte chemokine receptors by opioid receptors. *J Biol Chem* 278(15): 12729–12736
22 Rogers TJ, Peterson PK (2003) Opioid G protein-coupled receptors: signals at the crossroads of inflammation. *Trends Immunol* 4(3): 116–121
23 Bryant HU, Bernton EW, Kenner JR, Holaday JW (1991) Role of adrenal cortical activation in the immunosuppressive effects of chronic morphine treatment. *Endocrinology* 128(6): 3253–3258
24 Wang J, Charboneau R, Balasubramanian S, Barke RA, Loh HH, Roy S (2002) The immunosuppressive effects of chronic morphine treatment are partially dependent on corticosterone and mediated by the mu-opioid receptor. *J Leukoc Biol* 71(5): 782–790

25 Flores LR, Dretchen KL, Bayer BM (1996) Potential role of the autonomic nervous system in the immunosuppressive effects of acute morphine administration. *Eur J Pharmacol* 318(2–3): 437–446

26 Fecho K, Maslonek KA, Dykstra LA, Lysle DT (1993) Alterations of immune status induced by the sympathetic nervous system: immunomodulatory effects of DMPP alone and in combination with morphine. *Brain Behav Immun* 7(3): 253–270

27 Pitcher JA, Freedman NJ, Lefkowitz RJ (1998) G protein-coupled receptor kinases. *Annu Rev Biochem* 67: 653–692

28 Ali H, Richardson RM, Haribabu B, Snyderman R (1999) Chemoattractant receptor cross-desensitization. *J Biol Chem* 274(10): 6027–6030

29 Parent CA (2004) Making all the right moves: chemotaxis in neutrophils and Dictyostelium. *Curr Opin Cell Biol* 16(1): 4–13

30 McCarthy L, Wetzel M, Sliker JK, Eisenstein TK, Rogers TJ (2001) Opioids, opioid receptors, and the immune response. *Drug Alcohol Depend* 62(2): 111–123

31 Law PY, Wong YH, Loh HH (2000) Molecular mechanisms and regulation of opioid receptor signaling. *Annu Rev Pharmacol Toxicol* 40: 389–430

32 Mochly-Rosen D, Gordon AS (1998) Anchoring proteins for protein kinase C: a means for isozyme selectivity. *FASEB J* 12(1): 35–42

33 Kanzaki M, Mora S, Hwang JB, Saltiel AR, Pessin JE (2004) Atypical protein kinase C (PKCzeta/lambda) is a convergent downstream target of the insulin-stimulated phosphatidylinositol 3-kinase and TC10 signaling pathways. *J Cell Biol* 164(2): 279–290

34 Tran PB, Miller RJ (2003) Chemokine receptors: signposts to brain development and disease. *Nat Rev Neurosci* 4(6): 444–455

35 Ragozzino D (2002) CXC chemokine receptors in the central nervous system: Role in cerebellar neuromodulation and development. *J Neurovirol* 8(6): 559–572

36 Zou YR, Kottmann AH, Kuroda M, Taniuchi I, Littman DR (1998) Function of the chemokine receptor CXCR4 in haematopoiesis and in cerebellar development. *Nature* 393(6685): 595–599

37 Tani M, Fuentes ME, Peterson JW, Trapp BD, Durham SK, Loy JK, Bravo R, Ransohoff RM, Lira SA (1996) Neutrophil infiltration, glial reaction, and neurological disease in transgenic mice expressing the chemokine N51/KC in oligodendrocytes. *J Clin Invest* 98(2): 529–539

38 Giovannelli A, Limatola C, Ragozzino D, Mileo AM, Ruggieri A, Ciotti MT, Mercanti D, Santoni A, Eusebi F (1998) CXC chemokines interleukin-8 (IL-8) and growth-related gene product alpha (GROalpha) modulate Purkinje neuron activity in mouse cerebellum. *J Neuroimmunol* 92(1–2): 122–132

39 Meucci O, Fatatis A, Simen AA, Bushell TJ, Gray PW, Miller RJ (1998) Chemokines regulate hippocampal neuronal signaling and gp120 neurotoxicity. *Proc Natl Acad Sci USA* 95(24): 14500–14505

40 Oh SB, Tran PB, Gillard SE, Hurley RW, Hammond DL, Miller RJ (2001) Chemokines and glycoprotein120 produce pain hypersensitivity by directly exciting primary nociceptive neurons. *J Neurosci* 21(14): 5027–5035

41 Caterina MJ, Julius D (2001) The vanilloid receptor: a molecular gateway to the pain pathway. *Annu Rev Neurosci* 24: 487–517
42 Caterina MJ, Leffler A, Malmberg AB, Martin WJ, Trafton J, Petersen-Zeitz KR, Koltzenburg M, Basbaum AI, Julius D (2000) Impaired nociception and pain sensation in mice lacking the capsaicin receptor. *Science* 288(5464): 306–313
43 Shin J, Cho H, Hwang SW, Jung J, Shin CY, Lee SY, Kim SH, Lee MG, Choi YH, Kim J et al (2002) Bradykinin-12-lipoxygenase-VR1 signaling pathway for inflammatory hyperalgesia. *Proc Natl Acad Sci USA* 99(15): 10150–10155
44 Ji RR, Samad TA, Jin SX, Schmoll R, Woolf CJ (2002) p38 MAPK activation by NGF in primary sensory neurons after inflammation increases TRPV1 levels and maintains heat hyperalgesia. *Neuron* 36(1): 57–68
45 Chuang HH, Prescott ED, Kong H, Shields S, Jordt SE, Basbaum AI, Chao MV, Julius D (2001) Bradykinin and nerve growth factor release the capsaicin receptor from PtdIns(4,5)P2-mediated inhibition. *Nature* 411(6840): 957–962
46 Hu HJ, Bhave G, Gereau RW 4th (2002) Prostaglandin and protein kinase A-dependent modulation of vanilloid receptor function by metabotropic glutamate receptor 5: potential mechanism for thermal hyperalgesia. *J Neurosci* 22(17): 7444–7452
47 Sancho R, Lucena C, Macho A, Calzado MA, Blanco-Molina M, Minassi A, Appendino G, Munoz E (2002) Immunosuppressive activity of capsaicinoids: capsiate derived from sweet peppers inhibits NF-kappaB activation and is a potent antiinflammatory compound *in vivo*. *Eur J Immunol* 32(6): 1753–1763

Antimicrobial and related activities of chemokines

Osamu Yoshie

Department of Microbiology, Kinki University School of Medicine, Osaka-Sayama, Osaka 589-8511, Japan

Introduction

Chemokines are now known to play pivotal roles in both innate and acquired immunity primarily through their chemotactic activity for various leukocyte classes and subsets [1]. The family of antimicrobial peptides, also called natural antibiotics, constitutes the important immediate effector molecules against invading microorganisms [2, 3]. Accumulating evidence has revealed that the families of chemokines and antimicrobial peptides have substantially overlapping functions. While a number of antimicrobial peptides are chemotactic for selected classes and subsets of leukocyte [4], many chemokines have a substantial microbicidal activity against a broad spectrum of microorganisms [5–7]. Furthermore, CXCL16, a transmembrane-type chemokine [8, 9], was originally identified as a scavenger receptor termed SR-PSOX (scavenger receptor that binds phosphatidylserine and oxidized lipoprotein) [10]. Subsequently, a number of chemokines have been shown to display a similar binding activity for typical scavenger receptor ligands including oxidized lipoprotein and bacteria [11]. Thus, the family of chemokines may have substantial functional overlaps with the families of antimicrobial peptides and scavenger receptors. The overlapping functions of these distinct molecular families may have an evolutionary basis stemming from an ancient mode of recognition of pathogens and may represent a certain aspect of the pattern recognition of innate immunity.

The world of antimicrobial peptides

Antimicrobial peptides, now known by >700 in number, are the diverse family of small, mostly cationic polypeptides that have a direct killing activity against bacteria, fungi, parasite, and even some enveloped viruses [2, 3]. Peptides with similar structures and functions are found in virtually all branches of multicellular organisms. Their phylogenic relationships are, however, mostly unclear. This is mostly

Chemokine Biology – Basic Research and Clinical Application, Volume I
edited by Bernhard Moser, Gordon L. Letts and Kuldeep Neote

because there has been a strong evolutionary pressure for their gene multiplication and amino acid sequence diversification in order to cope with a wide variety of microorganisms [12–15]. The fundamental structural principal common to most antimicrobial peptides is the topological (rather than linear) amphipathic design, where clusters of hydrophobic and cationic amino acids are organized in discrete surface areas (Fig. 1). It is considered that the amphipathic and highly cationic nature of these peptides allows their selective binding and subsequent disruption of bacterial plasma membrane, which is much more negatively charged than that of host cells [2, 3]. Because of such an electrostatic and physicochemical mode of action, most antimicrobial peptides are only effective at relatively high (micomolar) concentrations and at low salt conditions [2, 3]. In mammals, therefore, the antimicrobial peptides are primarily involved in the barrier protection of various epithelial surfaces that are covered with a low salt body fluid. Some peptides are also involved in the non-oxidative bactericidal activity of leukocytes [2, 3].

For example, Paneth cells, which are present at the bottom of crypts in the small intestine, contain numerous large secretory granules that are discharged into the lumen upon various stimulations. Many components of these granules have potent antimicrobial properties and are likely to protect small intestine from microbial infection and colonization [16]. Paneth cells in humans express only two α-defensins, while mouse Paneth cells express not only more than 20 different α-defensins (also called as cryptdins) but also as many as 7 cryptdin-related sequence (CRS) peptides [13, 17]. CRS peptides represent a family of covalently linked homo- and hetero-dimeric antimicrobial molecules, a feature that may further contribute to their diversity for efficient protection of the gastrointestinal mucosa against enteropathogenic microorganisms [13]. Likewise, the non-oxidative mechanisms of human neutrophils are mediated by antimicrobial peptides and proteins stored within its various cytoplasmic granules [18, 19]. Cathepsin G, azurocidin (also called CAP37), BPI (also called CAP57), and α-defensins are restricted to the primary (asurophil) granules, which also contain myeloperoxidase, elastase, and proteinase 3 [18, 19]. Lactoferrin and hCAP-18 (the precursor of LL-37) are restricted to the neutrophil's secondary (specific) granules [18, 19]. Lysozyme, another antimicrobial molecule, occurs in both primary and secondary granules [18, 19]. Whereas azurophil granule contents are delivered preferentially to intracellular phagolysosomes, the specific granule contents are largely secreted extracellularly [18, 19]. Antimicrobial activity is also detected in natural killer cells and T cells, but the effector molecules that mediate the activity have not been systematically characterized. However, one effector molecule is granulysin, which has been shown to kill Gram-negative bacteria, Gram-positive bacteria, fungi and intracellular *Mycobacterium tuberculosis* [20]. Human cathelicidin LL-37 and α-defensins HNP 1–3 can be additional effector molecules for microbicidal activity of lymphocytes [21].

There is now substantial evidence that supports the vital role of the antimicrobial peptides in the host defense against bacterial infection (Tab. 1). For example,

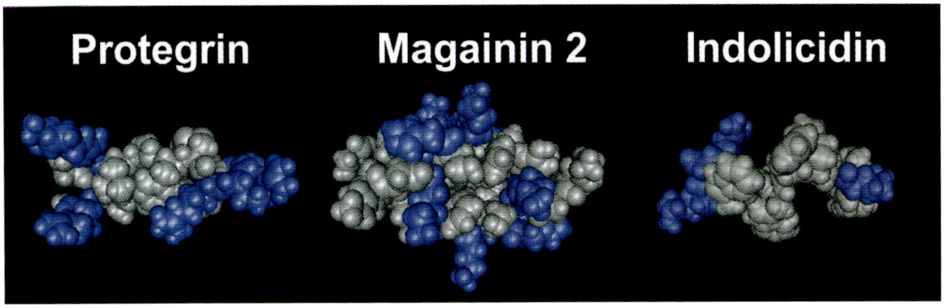

Figure 1
Topological clustering of catinonic and hydrophobic amino acids in antimicrobial peptides. Blue, basic (positively charged) amino acids; gray, hydrophobic amino acids.

the recurrent bacterial infection of lung in patients with cystic fibrosis could be in part due to poor performance of peptide-dependent antibacterial activity in the high-salt bronchotracheal fluid of these patients [22]. The abnormal expression of α-defensins and LL-37 correlates with the occurrence of severe periodontal infectious disease in patients with morbus Kostmann [23]. Mice deficient in the matalloprotease matrilysin, which is necessary to cleave the proforms of epithelial α-defensins in the small intestine, were shown to be more sensitive to orally administered bacteria [24]. Mice with targeted disruption of the cathelicidin gene *Cnlp* displayed a highly elevated susceptibility to Group A *Streptococcus* in a necrotizing cutaneous infection model [25]. Conversely, cathelicidin-resistant mutants of Group A *Streptococcus* demonstrated increased virulence *in vivo*, generating skin lesions of larger size and longer duration in wild-type mice [25]. Importantly, leukocytes derived from cathelicidin-deficient mice were functionally competent in chemotaxis and oxidative burst activity [25]. Thus, the absence of the antimicrobial peptide in the neutrophil granule and epidermal keratinocytes greatly compromises the host innate immunity against Group A *Streptococcus* infection [25]. Collectively, there is now little doubt about the vital role of antimicrobial peptides in innate immunity against invading microorganisms.

Chemotactic activity of antimicrobial peptides

Mammalian defensins and cathelicidins have also been shown to have multiple receptor-mediated effects on immune cells [4]. Most notably, many of them are chemotactic for selective leukocytes and apparently interact with pertussis toxin-sensitive Gαi-coupled seven-transmembrane receptors [4]. In this context, Yang et

Table 1 - In vivo *evidence for the vital role of antimicrobial peptides in host defense against bacterial infection*

Disease or genetic modification	Manifestation	Cause or mechanism	Refs.
cystic fibrosis	recurrent bacterial infection of the lung	high-salt inactivation of peptide-dependent antimicrobial activity	[22]
morbus Kostmann	severe periodontitis	lack of secretion of LL-37 in saliva	[23]
MMP-7 deficient mice	elevated susceptibility to orally administered bacteria	lack of processing of epithelial α-defensins	[24]
cathelicidin-deficient mice	elevated susceptibility to Group A *Streptococcus* skin infection	lack of cathelicidin expression in neutrophils and epithelial cells	[25]
β-defensin 1-deficient mice	poor clearance of *Haemophillus influenzae* in the lung colonization by *Staphylococcus* in the bladder	lack of α-defensin 1 expression in epithelial cells	[51, 52]
human α-defensin-5 transgenic mice	resistance to oral challenge with *Salmonella syphimurium*	transgenic expression of human α-defensin 5 in Paneth cells	[53]

al. have demonstrated that human β-defensins are potent agonists for CCR6 [26], the receptor for a chemokine CCL20/LARC, which is expressed by various epithelial cells and attracts immature dendritic cells and effector lymphocytes [1, 27–30]. In fact, β-defensins appear to have a tertiary structure very similar to that of CCL20 and thus may act as "minichemokines" [31]. Furthermore, LL-37 has been shown to attract neutrophils, monocytes, and mast cells via human formyl peptide receptor-like 1 (FPRL1) [32]. Its angiogenic activity is also mediated by FPRL1 expressed on endothelial cells [26]. While human β-defensins HBD1–3 and mouse β-defensins mBD2 and 3 attract immature dendritic cells via CCR6, HBD3 may also use a receptor other than CCR6 for attraction of monocytes because these cells do not express CCR6 [4]. Human α-defensins HNP1–3 also use an unknown Gαi-protein-coupled receptor(s) because their chemotactic activity can be blocked by pretreatment of target cells with pertussis toxin [4]. Collectively, it is now clear that many antimicrobial peptides can be regarded as endogenous ligands for some Gαi-protein-coupled chemotactic receptors. Thus, besides direct killing of invading microorganisms, antimicrobial peptides may also have an important role in the recruitment of leukocytes in innate and acquired immunity.

Antimicrobial activity of chemokines

Chemokines play pivotal roles in both innate and acquired immunity primarily by inducing directed migration of various leukocyte classes and subsets via interactions with a group of Gαi-protein-coupled seven transmembrane receptors [1]. Furthermore, recent studies have revealed that many chemokines have a direct microbicidal activity (Tab. 2). Krijgsveld et al. determined the amino acid sequences of the purified antibacterial molecules termed thrombocidins that were stored in the α-granules of human platelets [33]. The molecules turned out to be two related chemokine variants processed from a common precursor platelet basic protein (PBP) and truncated by two amino acids in the C terminus, namely, NAP-2/CXCL7(59–126) and CTAP-III/CXCL7(44–126) [33]. The full-length NAP-2/CXCL7(59–128) and CTAP-III/CXCL7(44–128) were not microbicidal in their hands [33]. Tang et al. also characterized antimicrobial molecules released by human platelets upon thrombin stimulation [34]. They demonstrated that several platelet chemokines including CXCL4/PF-4, CCL5/RANTES, the full-length CTAP-III/CXCL7(44–128) and the CTAP-III precursor PBP/CXCL7(35–128) had potent antimicrobial activity against Gram-negative *Escherichia coli*, Gram-positive *Staphylococcus aureus*, *Cryptococcus neoformans*, and, with the exception of CTAP-III and PBP, *Candida albicans* [34]. In their hands, thus, the full-length CTAP-III was also active. Furthermore, Cole et al. examined a panel of 11 chemokines representing all four chemokine subfamilies for antimicrobial activity and demonstrated that the three IFN-inducible non-ELR-motif CXC chemokines, MIG/CXCL9, IP-10/CXCL10, and I-TAC/CXCL11, were microbicidal against *Escherichia coli* and Gram-positive *Listeria monocytogenes* [5]. We also reported a broad-spectrum antimicrobial activity of CCL28/MEC (see below) [6], a chemokine selectively expressed by various mucosal tissues [35, 36]. Yang et al., who have originally reported that human β-defensins are functional ligands for CCR6 [26], also tested whether CCL20/LARC was in converse microbicidal [7]. They found that, similar to β-defensins, CCL20 was microbicidal against *Escherichia coli*, *Pseudomonas aeruginosa*, *Moraxella catarrhalis*, *Streptococcus pyogenes*, *Enterococcus faecium*, *Staphylococcus aureus*, and *Candida albicans* [7]. Furthermore, they demonstrated that many other chemokines also displayed similar antimicrobial activities [7]. These included CXCL1/Gro-α, CXCL2/Gro-β, CXCL3/Gro-γ, CXCL12/SDF-1, CXCL13/BLC, CXCL14/BRAK, CCL1/I-309, CCL8//MCP-2, CCL11/Eotaxin, CCL13/MCP-4, CCL17/TARC, CCL18/PARC, CCL19/ELC, CCL21/SLC, CCL22/MDC, CCL25/ TECK, and XCL1/Lymphotactin [7]. Thus, about two-thirds of the chemokines that were investigated in their study showed the capacity to kill microorganisms *in vitro*. Most bactericidal chemokines, in particular CXCL1, CXCL2, CXCL3, CXCL12, CXCL13, CCL1, CCL13, CCL19, CCL20, and XCL1, were more potent against Gram-negative *E. coli* than against Gram-positive *S. aureus*. A striking difference was observed between the antimicrobial activity of closely related CCL19 and CCL21 [1]. CCL19

Table 2 - Evidence for antimicrobial activity of chemokines

Authors	Source	Target microorganisms	Active chemokines	Inactive chemokines	Refs.
Kriejgsved et al.	Human platelets	*B. subsitis*, *E. coli*, *S. aureus*, *L. lactis*, *C. neoformans*	NAP-2/CXCL7(59-126) CTAP-III/CXCL7(44-126)	Full-length NAP-2(59-128) Full-length CTAP-III(44-128)	[33]
Tang et al.	Human platelets	*E. coli*, *S. aureus*, *C. albicans*, *C. neoformans*	PF-4/CXCL4, RANTES/CCL5, CTAP-III/CXCL7(44-128) PBP/CXCL7(35-128)		[34]
Cole et al.	Recombinant proteins	*E. coli*, *L. monocytogenes*	MIG/CXCL9, IP-10/CXCL10, I-TAC/CXCL11	IL-8/CXCL8, ENA-78/CXCL5, MCP-1/CCL2, MIP-1α/CCL3, MIP-1β/CCL4, RANTES/CCL5, FTN/CX3CL1, LTN/XCL1	[5]
Hoover et al.	Synthetic proteins	*E. coli*, *S. aureus*, *C. albicans*	LARC/CCL20	MCP-1/CCL2	[54]
Hieshima et al.	Recombinant proteins	*C. albicans*, *P. aeruginosa*, *K. pneumoniae*, *S. mutans*, *S. pyogenes*, *S. aureus*	MEC/CCL28	CTACK/CCL27	[6]
	Recombinant proteins	*E. coli*, *P. aeruginosa*, *M. catarrhalis*, *S. pyogenes*, *E. faecium*, *S. aureus*, *C. albicans*	Groα/CXCL1, Groβ/CXCL2, Groγ/CXCL3, SDF-1/CXCL12, BLC/CXCL13, BRAK/CXCL14, I-309/CCL1, MCP-2/CCL8, eotaxin/CCL11, MCP-4/CCL13, TARC/CCL17, PARC/CCL18, ELC/CCL19, LARC/CCL20, SLC/CCL21, MDC/CCL22, TECK/CCL25, LTN/XCL1	GCP-2/CXCL6, IL-8/CXCL8, MCP-1/CCL2, MIP-1α/CCL3, RANTES/CCL5, MCP-3/CCL7, LEC/CCL16, CTACK/CCL27, FTN/CX3CL1	[7]

was active against *E. coli* with little detectable activity against *S. aureus*. On the other hand, CCL21 demonstrated a potent activity against *S. aureus*, while being less potent against *E. coli* than CCL19 [7]. Even though there are some discrepancies concerning antimicrobial activity of some chemokines (Tab. 2), these studies have clearly demonstrated that many chemokines have an intrinsic microbicidal activity when tested in low salt assay conditions *in vitro*.

In particular, CCL28/MEC is expressed at high levels in the mucosal tissues such as salivary glands, trachea, colon, and mammary glands [35, 36]. CCL28 is most homologous with CCL27/CTACK, which is selectively expressed in the skin [37, 38]. These two chemokines commonly act on CCR10 [35, 36, 39, 40]. We observed that CCL28 was not only strongly expressed in the salivary glands but also secreted into the saliva and milk at relatively high concentrations [6]. Furthermore, we noticed that the extended C-terminal regions of CCL28 is highly enriched with histidine residues and shows a significant sequence similarity with histatin-5, a histidine-rich candidacidal peptide secreted in human saliva [6, 41]. These observations led us to examine potential microbicidal activity of CCL28 and its C-terminal peptide. As summarized in Table 3, we found that CCL28 indeed exerts a potent antimicrobial activity against not only *Candida albicans*, but also against Gram-negative bacteria and Gram-positive bacteria [6]. Like histatin-5, the synthetic peptide corresponding to the 28-amino acid C-terminal segment of CCL28 (CCL28-C) also showed a selective antimicrobial activity against *C. albicans* [6]. On the other hand, CCL27, which is most closely related to CCL28 [37, 38], hardly showed such antimicrobial activity [6]. CCL28 rapidly generated pores in the membrane of target microbes [6]. Like many other antimicrobial chemokines and peptides, the microbicidal activity of CCL28 is salt-sensitive [6]. In this context, it should be noted that the mucosal fluids such as saliva, milk, and tracheal and colonic secretions are commonly low in salt concentrations. Thus, CCL28, which is secreted into low-salt body fluids at high concentrations, may have a potential as a direct microbicidal factor. It is also noteworthy that the chemokines with potent antimicrobial activities such as CXCL9, CXCL10, CXCL11, and CCL20 are all expressed and secreted at relatively high concentrations by various epithelial cells [27–30, 42]. Collectively, some chemokines may have a substantial role in host defense against microorganisms as direct microbicidal agents.

Common structural features of chemokines with antimicrobial activity

Like many other antimicrobial peptides, the chemokines with antimicrobial activity tend to have a higher pI than those without such activity, indicating that cationicity is an important feature for antimicrobial chemokines [7]. However, cationicity alone is not sufficient to distinguish chemokines with and without antimicrobial activity. Furthermore, the potency of antimicrobial chemokines does not directly

Table 3 - Summary of antimicrobial activity of CCL28

Microbe	IC$_{50}$ (μM)				
	CCL28	mCCL28	CCL27	CCL28-C	Histatin-5
P. aeruginosa	0.4 ± 0.1	1.7 ± 0.1	>10	>10	>10
K. pneumoniae	0.3 ± 0.1	1.6 ± 0.1	>10	>10	3.0 ± 0.7
S. mutans	1.7 ± 0.4	1.5 ± 0.3	>10	>10	>10
S. pyogenes	3.0 ± 0.2	4.5 ± 0.4	>10	>10	>10
S. aureus	0.9 ± 0.1	0.9 ± 0.1	>10	7.0 ± 1.2	>10
C. albicans	0.7 ± 0.2	1.3 ± 1.0	5.0 ± 1.9	1.6 ± 0.4	3.5 ± 1.6

IC$_{50}$, 50% inhibitory concentration; mCCL28, mouse CCL28; CCL28-C, the C-terminal 28 amino acid peptide of CCL28

correlate with their cationicity. Therefore, in addition to cationicity, other structural features are necessary for a given chemokine to have an antimicrobial activity [7]. As shown in Figure 2, comparison of the structures between chemokines with and without antimicrobial activities suggests that the topological formation of a large, positively charged electrostatic patch on the surface of the molecule is likely to be a common feature of antimicrobial chemokines. The rest of the molecule is mostly hydrophobic with spotted negative electrostatic charges.

Scavenger receptor activity of chemokines

Scavenger receptors are a highly heterogeneous group of cell surface molecules that commonly bind and internalize oxidized low density lipoprotein (OxLDL) and polyanionic molecules [43]. Scavenger receptors are expressed by myeloid cells (macrophages and dendritic cells) and some endothelial cells, and play an important role in uptake and clearance of modified host molecules, apoptotic cells, microorganisms, and their products [44]. CXCL16, a transmembrane-type chemokine [8, 9], was originally identified as a scavenger receptor for oxidized lipoprotein [10]. CXCL16 is expressed by cells such as macrophages and dendritic cells, and has been shown to bind and internalize various scavenger receptor ligands such as oxidized lipoprotein, bacteria, and sulfated polyanions [10, 45]. Shimaoka et al. have shown that not only CXCL16, but also 12 out of 15 chemokines examined are capable of binding typical scavenger receptor ligands such as OxLDL, Gram-positive bacteria, and Gram-negative bacteria [11]. Furthermore, OxLDL effectively blocks the binding of such chemokines to their respective receptors, suggesting that the receptor binding site of these chemokines mostly overlaps with their potential binding site for

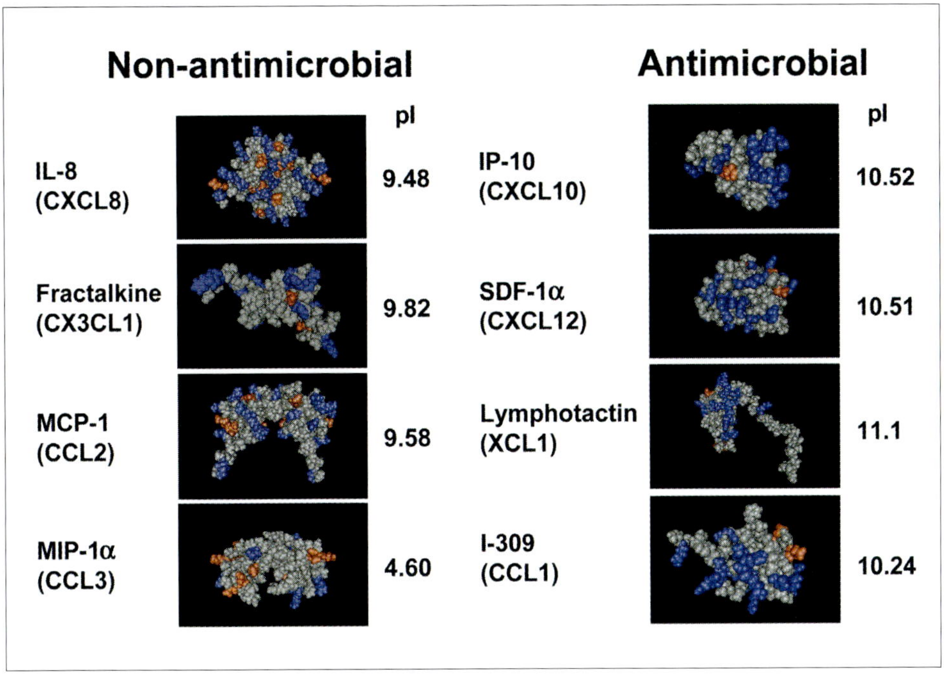

Figure 2
Topological distribution of charged amino acids in non-antimicrobial and antimicrobial chemokines
The pI value of each chemokine is indicated on the right. Red, acidic (negatively charged); blue, basic (positively charged); grey, hydrophobic/neutral.

OxLDL [11]. Indeed, both the chemotactic and scavenger receptor activities of CXCL16 were similarly impaired by a series of mutations in the chemokine domain [11]. As expected, the chemokines with antimicrobial activity consistently bound more avidly with OxLDL and bacteria than those without antimicrobial activity [11].

Concluding remarks

It is now apparent that many chemokines have a potential antimicrobial activity and can also avidly bind OxLDL and other scavenger receptor ligands including bacteria. Thus, chemokines, antimicrobial peptides, and scavenger receptors have some molecular properties in common. The evolutionary origin of such shared properties is not clear but could be related to an ancient pattern recognition of microbial pathogens by the host [46]. Alternatively, such properties might have been acquired through evolutionary conversion. At any rate, there could have been a strong selective pressure toward retaining and/or acquiring some common molecular features.

The obvious common property of chemokines with antimicrobial peptides and scavenger receptors is cationicity. This could be essential for the antimicrobial peptides and scavenger receptors to recognize bacterial cells that have much higher negative charges than host cells [2, 3]. On the other hand, there may not be such intrinsic functional necessity for chemokine *per se* to be cationic. However, one important reason for most chemokines to be cationic is that the N-terminal regions of the chemokine receptors are highly rich in acidic residues and even sulfated at some tyrosine residues [47, 48]. In fact, many chemoattractant receptors are commonly negatively charged at their N-terminal extracellular domains [48]. Currently, most chemokines are considered to interact with their receptors in a two-step process [49]. The first high-affinity interaction mainly involves the N-terminal region of the receptor and is mostly mediated by strong electrostatic force. The subsequent lower affinity interaction involves other extracellular loops of receptors, while the N-terminal region of chemokines plays a critical role in signaling. Chemokines also interact with negatively charged glycosaminoglycans such as heparin and heparan sulfate, and this property is necessary for their *in vivo* activity [50]. These biological requirements may in part explain the common cationic property of most chemokines. Thus, their possession of antimicrobial and scavenger receptor-like activities may be mostly fortuitous (a matter of *in vitro* assays) but may still have some physiologic implications for some chemokines.

At present, the antimicrobial activity of chemokines has been shown only by *in vitro* assays. Thus, studies using knockout mice or transgenic mice would be necessary to prove any physiologic role of chemokines in direct microbial killing *in vivo*. Given the micromolar concentrations required for effective microbicidal activity, however, it is unlikely that direct killing of microorganisms is a major function of any chemokines. However, still some chemokines may play a significant role in direct killing of microorganisms through cooperation with other chemokines and other antimicrobial peptides. In contrast, the chemotactic activity of antimicrobial peptides are more physiologically attainable, requiring only nanomolar concentrations [4]. Furthermore, there could still be a large number of new antimicrobial peptides that remain to be characterized. For example, an improved genome-wide search has recently identified a total of 28 new human and 43 new mouse β-defensin genes that are clustered in five syntenic chromosomal regions [15]. Thus, it is quite a challenge to characterize such new peptides for their antimicrobial spectrum and chemotactic activity, and to identify their chemotactic receptors.

Acknowledgements
I thank the help of K. Hieshima and T. Nakayama for writing this article.

References

1 Yoshie O, Imai T, Nomiyama H (2001) Chemokines in immunity. *Adv Immunol* 78: 57–110

2 Ganz T (2003) Defensins: antimicrobial peptides of innate immunity. *Nat Rev Immunol* 3: 710–720

3 Zasloff M (2002) Antimicrobial peptides of multicellular organisms. *Nature* 415: 389–395

4 Yang D, Biragyn A, Hoover DM, Lubkowski J, Oppenheim JJ (2004) Multiple roles of antimicrobial defensins, cathelicidins, and eosinophil-derived neurotoxin in host defense. *Annu Rev Immunol* 22: 181–215

5 Cole AM, Ganz T, Liese AM, Burdick MD, Liu L, Strieter RM (2001) Cutting edge: IFN-inducible ELR-CXC chemokines display defensin-like antimicrobial activity. *J Immunol* 167: 623–627

6 Hieshima K, Ohtani H, Shibano M, Izawa D, Nakayama T, Kawasaki Y, Shiba F, Shiota M, Katou F, Saito T et al (2003) CCL28 has dual roles in mucosal immunity as a chemokine with broad-spectrum antimicrobial activity. *J Immunol* 170: 1452–1461

7 Yang D, Chen Q, Hoover DM, Staley P, Tucker KD, Lubkowski J, Oppenheim JJ (2003) Many chemokines including CCL20/MIP-3alpha display antimicrobial activity. *J Leukoc Biol* 74: 448–455

8 Matloubian M, David A, Engel S, Ryan JE, Cyster JG (2000) A transmembrane CXC chemokine is a ligand for HIV-coreceptor Bonzo. *Nat Immunol* 1: 298–304

9 Wilbanks A, Zondlo SC, Murphy K, Mak S, Soler D, Langdon P, Andrew DP, Wu L, Briskin M (2001) Expression cloning of the STRL33/BONZO/TYMSTRligand reveals elements of CC, CXC, and CX3C chemokines. *J Immunol* 166: 5145–5154

10 Shimaoka T, Kume N, Minami M, Hayashida K, Kataoka H, Kita T, Yonehara S (2000) Molecular cloning of a novel scavenger receptor for oxidized low density lipoprotein, SR-PSOX, on macrophages. *J Biol Chem* 275: 40663–40666

11 Shimaoka T, Nakayama T, Hieshima K, Kume N, Fukumoto N, Minami M, Hayashida K, Kita T, Yoshie O, Yonehara S (2004) Chemokines generally exhibit scavenger receptor activity through their receptor-binding domain. *J Biol Chem* 279: 26807–26810

12 Domachowske JB, Bonville CA, Dyer KD, Rosenberg HF (1998) Evolution of antiviral activity in the ribonuclease A gene superfamily: evidence for a specific interaction between eosinophil-derived neurotoxin (EDN/RNase 2) and respiratory syncytial virus. *Nucleic Acids Res* 26: 5327–5332

13 Hornef MW, Putsep K, Karlsson J, Refai E, Andersson M (2004) Increased diversity of intestinal antimicrobial peptides by covalent dimer formation. *Nat Immunol* 5: 836–843

14 Hughes AL (1999) Evolutionary diversification of the mammalian defensins. *Cell Mol Life Sci* 56: 94–103

15 Schutte BC, Mitros JP, Bartlett JA, Walters JD, Jia HP, Welsh MJ, Casavant TL, McCray PB Jr (2002) Discovery of five conserved beta-defensin gene clusters using a computational search strategy. *Proc Natl Acad Sci USA* 99: 2129–2133

16 Ayabe T, Ashida T, Kohgo Y, Kono T (2004) The role of Paneth cells and their antimicrobial peptides in innate host defense. *Trends Microbiol* 12: 394–398
17 Cunliffe RN (2003) Alpha-defensins in the gastrointestinal tract. *Mol Immunol* 40: 463–467
18 Borregaard N, Cowland JB (1997) Granules of the human neutrophilic polymorphonuclear leukocyte. *Blood* 89: 3503–3521
19 Chertov O, Yang D, Howard OM, Oppenheim JJ (2000) Leukocyte granule proteins mobilize innate host defenses and adaptive immune responses. *Immunol Rev* 177: 68–78
20 Krensky AM (2000) Granulysin: a novel antimicrobial peptide of cytolytic T lymphocytes and natural killer cells. *Biochem Pharmacol* 59: 317–320
21 Agerberth B, Charo J, Werr J, Olsson B, Idali F, Lindbom L, Kiessling R, Jornvall H, Wigzell H, Gudmundsson GH (2000) The human antimicrobial and chemotactic peptides LL-37 and alpha-defensins are expressed by specific lymphocyte and monocyte populations. *Blood* 96: 3086–3093
22 Smith JJ, Travis SM, Greenberg EP, Welsh MJ (1996) Cystic fibrosis airway epithelia fail to kill bacteria because of abnormal airway surface fluid. *Cell* 85: 229–236
23 Putsep K, Carlsson G, Boman HG, Andersson M (2002) Deficiency of antibacterial peptides in patients with morbus Kostmann: an observation study. *Lancet* 360: 1144–1149
24 Wilson CL, Ouellette AJ, Satchell DP, Ayabe T, Lopez-Boado YS, Stratman JL, Hultgren SJ, Matrisian LM, Parks WC (1999) Regulation of intestinal alpha-defensin activation by the metalloproteinase matrilysin in innate host defense. *Science* 286: 113–117
25 Nizet V, Ohtake T, Lauth X, Trowbridge J, Rudisill J, Dorschner RA, Pestonjamasp V, Piraino J, Huttner K, Gallo RL (2001) Innate antimicrobial peptide protects the skin from invasive bacterial infection. *Nature* 414: 454–457
26 Yang D, Chertov O, Bykovskaia SN, Chen Q, Buffo MJ, Shogan J, Anderson M, Schroder JM, Wang JM, Howard OM et al (1999) Beta-defensins: linking innate and adaptive immunity through dendritic and T cell CCR6. *Science* 286: 525–528
27 Fujiie S, Hieshima K, Izawa D, Nakayama T, Fujisawa R, Ohyanagi H, Yoshie O (2001) Proinflammatory cytokines induce liver and activation-regulated chemokine/macrophage inflammatory protein-3alpha/CCL20 in mucosal epithelial cells through NF-kappaB [correction of NK-kappaB]. *Int Immunol* 13: 1255–1263
28 Nakayama T, Fujisawa R, Yamada H, Horikawa T, Kawasaki H, Hieshima K, Izawa D, Fujiie S, Tezuka T, Yoshie O (2001) Inducible expression of a CC chemokine liver- and activation-regulated chemokine (LARC)/macrophage inflammatory protein (MIP)-3 alpha/CCL20 by epidermal keratinocytes and its role in atopic dermatitis. *Int Immunol* 13: 95–103
29 Shirane J, Nakayama T, Nagakubo D, Izawa D, Hieshima K, Shimomura Y, Yoshie O (2004) Corneal epithelial cells and stromal keratocytes efficiently produce CC chemokine-ligand 20 (CCL20) and attract cells expressing its receptor CCR6 in mouse herpetic stromal keratitis. *Curr Eye Res* 28: 297–306
30 Starner TD, Barker CK, Jia HP, Kang Y, McCray BP Jr (2003) CCL20 is an inducible

product of human airway epithelia with innate immune properties. *Am J Respir Cell Mol Biol* 29: 627–633

31 Perez-Canadillas JM, Zaballos A, Gutierrez J, Varona R, Roncal F, Albar JP, Marquez G, Bruix M (2001) NMR solution structure of murine CCL20/MIP-3alpha, a chemokine that specifically chemoattracts immature dendritic cells and lymphocytes through its highly specific interaction with the beta-chemokine receptor CCR6. *J Biol Chem* 276: 28372–28379

32 De Y, Chen Q, Schmidt AP, Anderson GM, Wang JM, Wooters J, Oppenheim JJ, Chertov O (2000) LL-37, the neutrophil granule- and epithelial cell-derived cathelicidin, utilizes formyl peptide receptor-like 1 (FPRL1) as a receptor to chemoattract human peripheral blood neutrophils, monocytes, and T cells. *J Exp Med* 192: 1069–1074

33 Krijgsveld J, Zaat SA, Meeldijk J, van Veelen PA, Fang G, Poolman B, Brandt E, Ehlert JE, Kuijpers AJ, Engbers H et al (2000) Thrombocidins, microbicidal proteins from human blood platelets, are C-terminal deletion products of CXC chemokines. *J Biol Chem* 275: 20374–20381

34 Tang YQ, Yeaman MR, Selsted ME (2002) Antimicrobial peptides from human platelets. *Infect Immun* 70: 6524–6533

35 Pan J, Kunkel EJ, Gosslar U, Lazarus N, Langdon P, Broadwell K, Vierra MA, Genovese MC, Butcher EC, Soler D (2000) A novel chemokine ligand for CCR10 and CCR3 expressed by epithelial cells in mucosal tissues. *J Immunol* 165: 2943–2949

36 Wang W, Soto H, Oldham ER, Buchanan ME, Homey B, Catron D, Jenkins N, Copeland NG, Gilbert DJ, Nguyen N et al (2000) Identification of a novel chemokine (CCL28), which binds CCR10 (GPR2). *J Biol Chem* 275: 22313–22323

37 Ishikawa-Mochizuki I, Kitaura M, Baba M, Nakayama T, Izawa D, Imai T, Yamada H, Hieshima K, Suzuki R, Nomiyama H et al (1999) Molecular cloning of a novel CC chemokine, interleukin-11 receptor alpha-locus chemokine (ILC), which is located on chromosome 9p13 and a potential homologue of a CC chemokine encoded by molluscum contagiosum virus. *FEBS Lett* 460: 544–548

38 Morales J, Homey B, Vicari AP, Hudak S, Oldham E, Hedrick J, Orozco R, Copeland NG, Jenkins NA, McEvoy LM et al (1999) CTACK, a skin-associated chemokine that preferentially attracts skin-homing memory T cells. *Proc Natl Acad Sci USA* 96: 14470–14475

39 Homey B, Wang W, Soto H, Buchanan ME, Wiesenborn A, Catron D, Muller A, McClanahan TK, Dieu-Nosjean MC, Orozco R et al (2000) Cutting edge: the orphan chemokine receptor G protein-coupled receptor-2 (GPR-2, CCR10) binds the skin-associated chemokine CCL27 (CTACK/ALP/ILC). *J Immunol* 164: 3465–3470

40 Jarmin DI, Rits M, Bota D, Gerard NP, Graham GJ, Clark-Lewis I, Gerard C (2000) Cutting edge: identification of the orphan receptor G-protein-coupled receptor 2 as CCR10, a specific receptor for the chemokine ESkine. *J Immunol* 164: 3460–3464

41 Tsai H, Bobek LA (1998) Human salivary histatins: promising anti-fungal therapeutic agents. *Crit Rev Oral Biol Med* 9: 480–497

42 Sauty A, Dziejman M, Taha RA, Iarossi AS, Neote K, Garcia-Zepeda EA, Hamid Q,

Luster AD (1999) The T cell-specific CXC chemokines IP-10, Mig, and I-TAC are expressed by activated human bronchial epithelial cells. *J Immunol* 162: 3549–3558

43 Steinbrecher UP (1999) Receptors for oxidized low density lipoprotein. *Biochim Biophys Acta* 1436: 279–298

44 Peiser L, Mukhopadhyay S, Gordon S (2002) Scavenger receptors in innate immunity. *Curr Opin Immunol* 14: 123–128

45 Shimaoka T, Nakayama T, Kume N, Takahashi S, Yamaguchi J, Minami M, Hayashida K, Kita T, Ohsumi J, Yoshie O et al (2003) Cutting edge: SR-PSOX/CXC chemokine ligand 16 mediates bacterial phagocytosis by APCs through its chemokine domain. *J Immunol* 171: 1647–1651

46 Gordon S (2002) Pattern recognition receptors: doubling up for the innate immune response. *Cell* 111: 927–930

47 Farzan M, Mirzabekov T, Kolchinsky P, Wyatt R, Cayabyab M, Gerard NP, Gerard C, Sodroski J, Choe H (1999) Tyrosine sulfation of the amino terminus of CCR5 facilitates HIV-1 entry. *Cell* 96: 667–676

48 Murphy PM (1994) The molecular biology of leukocyte chemoattractant receptors. *Annu Rev Immunol* 12: 593–633

49 Monteclaro FS, Charo IF (1997) The amino-terminal domain of CCR2 is both necessary and sufficient for high affinity binding of monocyte chemoattractant protein 1. Receptor activation by a pseudo-tethered ligand. *J Biol Chem* 272: 23186–23190

50 Proudfoot AE, Handel TM, Johnson Z, Lau EK, LiWang P, Clark-Lewis I, Borlat F, Wells TN, Kosco-Vilbois MH (2003) Glycosaminoglycan binding and oligomerization are essential for the *in vivo* activity of certain chemokines. *Proc Natl Acad Sci USA* 100: 1885–1890

51 Morrison G, Kilanowski F, Davidson D, Dorin J (2002) Characterization of the mouse beta defensin 1, Defb1, mutant mouse model. *Infect Immun* 70: 3053–3060

52 Moser C, Weiner DJ, Lysenko E, Bals R, Weiser JN, Wilson JM (2002) beta-Defensin 1 contributes to pulmonary innate immunity in mice. *Infect Immun* 70: 3068–3072

53 Salzman NH, Ghosh D, Huttner KM, Paterson Y, Bevins CL (2003) Protection against enteric salmonellosis in transgenic mice expressing a human intestinal defensin. *Nature* 422: 522–526

54 Hoover DM, Boulegue C, Yang D, Oppenheim JJ, Tucker K, Lu W, Lubkowski J (2002) The structure of human macrophage inflammatory protein-3alpha /CCL20. Linking antimicrobial and CC chemokine receptor-6-binding activities with human beta-defensins. *J Biol Chem* 277: 37647–37654

Virus-encoded chemokine modulators as novel anti-inflammatory reagents

Alexandra Lucas[1,2,3], *Dana McIvor*[1,2,3] *and Grant McFadden*[1,3]

[1]BioTherapeutics Research Group; [2]Vascular Biology Research Group, Robarts Research Institute; [3]Dept. Microbiology and Immunology, University of Western Ontario, London, Ontario, Canada N6G 2K5

Viruses are potent manipulators of chemokines

Viruses that successfully invade immunocompetent hosts do so by acquiring self-protective strategies to evade or subvert the amalgamated forces of the innate and acquired immune responses [1–3]. Studies of individual viral anti-immune mechanisms tend to shed light on specific pathways that regulate the immune or inflammatory responses encountered by specific viruses within susceptible hosts. Viruses as a whole can express effector molecules that target the entire gamut of immune pathways of vertebrate hosts, but several pathways stand out as being particularly targeted by viruses from many distinct families. For example, a survey of the host antiviral response pathways already known to be targeted by viruses reveals a spectrum of key immune targets: targets such as antigen presentation, apoptosis, intracellular signaling, toll-like receptors, cytokines chemokines, serine proteinases, cytotoxic killing mechanisms, antibody generation, and humoral regulators, etc. [4–9]. In fact, the ever-growing collection of viral strategies that modulate the immune system can be considered as comprising the discipline of "anti-immunology" and is the subject of a vast body of scientific literature [10–16]. In particular, the chemokine circuitry has been frequently targeted by viruses for manipulation by three classes of virus-encoded regulators: (1) chemokine mimics, (2) chemokine binding proteins and (3) chemokine receptor homologs [17–20].

In many cases, viruses have evolved chemokine regulators that counteract the inflammatory responses of the host, thus endowing these molecules with highly specific anti-inflammatory properties [21–27]. Viruses do not generally express immunomodulators that require high concentration in order to effectively perturb their intended immune pathways. Rather, viruses have evolved to express host-directed regulators that can be delivered transiently at exceeding low dosages (less than nanomolar) within a selected microenvironment of the infected tissues. The combination of high potency and highly specific targeting provides a powerful

Chemokine Biology – Basic Research and Clinical Application, Volume I
edited by Bernhard Moser, Gordon L. Letts and Kuldeep Neote

platform with which to develop next-generation drugs based on viral protein immunomodulators to treat diseases associated with excessive inflammation [28].

In this chapter, we focus on virus-encoded chemokine modulators that are secreted from infected cells and target chemokines as competing ligands or as binding proteins (Fig. 1). In particular, we discuss in greater depth those virus-encoded chemokine regulators that have been tested individually, in the absence of virus infection, and examined as therapeutic reagents in models of diseases associated with excessive inflammation or immune responses (Tabs 1 and 2).

Chemokines – their role in the inflammatory response

Chemokines are small 8–12 kDa proteins that provide a chemoattractant function enticing circulating cells in the blood into sites of injury or infection [29–34]. The chemokines have been classified by arrangement of the N-terminal cysteine residues relative to one another, into C, CC, CXC and CX_3C classes, X representing amino acids inserted between the C amino acids [34]. The CC and CXC classes have been the most extensively studied, with CC chemokines having a proclivity toward attraction of monocytes and lymphocytes, but in reality the chemokines and their receptors are both redundant and promiscuous often crossing class activities and receptor affiliations [32–35]. The C-terminus of many chemokines recognize glycosaminoglycans (GAGs) and is thought to provide an anchor for chemokines to establish a solid phase gradient that can act to directionally attract cells into the target tissues [31, 36]. The N-terminus of chemokines recognize the appropriate seven transmembrane G-protein-coupled chemokine receptors present on the surface of the attracted leukocytes, thereby allowing the cells to become both adherent to the chemokine and also aiding in their activation [31, 34, 36]. Chemokine receptors can be classified in the same manner as chemokines (C, CC, CXC, CX3C) and are either inducible (inflammatory) or constitutively expressed [29, 35]. After chemokine binding, the subunits of the associated G protein dissociate from the chemokine receptor. The G beta and gamma subunits activate an assortment of enzymes while the G alpha subunit regulates production of cAMP or can couple chemokine receptor activation to non-receptor protein-tyrosine-kinase-initiated pathways [31, 37, 38]. Both chemokines and chemokine receptors have been implicated in the initiation and progression of many diseases, including: arthritis, infections such as AIDS, glomerulonephritis, neurotrauma, inflammatory CNS disorders, atherosclerosis, myocardial damage, lung diseases and transplant rejection, among many others [29–35, 37–39]. Figure 2 illustrates some of the numerous chemokine and receptor pathways associated with disease progression which have only been very briefly touched on in this chapter. This figure also illustrates the potential levels of inhibition by viral modulators: specifically, the modulation of chemokine gradient formation (blue circle) and ligand-receptor recognition (aqua circle). Viral chemokine

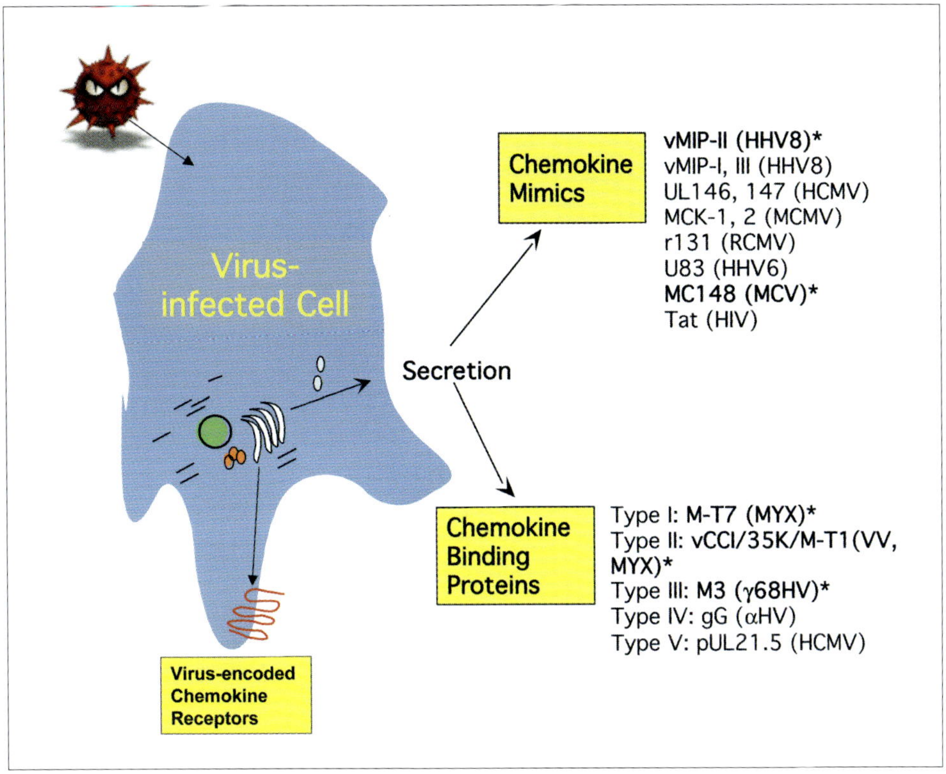

Figure 1
Viral chemokine modulators. The proteins marked with an asterix are considered in greater detail in this review.

modulators that target or bind to chemokines and their receptors thus have the potential to modify and even completely arrest chemokine mediated responses during the inflammatory system response.

Secreted immunomodulatory viral proteins as anti-inflammatory reagents

Virus-encoded immunomodulatory proteins have been identified from many virus families, with the majority being derived from DNA viruses that express multiple genes in addition to those required just for virus replication and propagation in tissue culture. Because of their large genomic sizes, members of the poxvirus and herpesvirus families have evolved to encode more such immunomodulators than other viruses [1, 10–16, 40–42]. In some cases, the origins of these immunomodulatory

Table 1 - Viral chemokine binding proteins tested in animal disease models

	Viral protein	Delivery mode	Animal model	Refs
(I)	M-T7 (Myxoma)	Protein (i.v.)	Vascular hyperplasia (rat, rabbit)	[75]
		Protein (i.v.)	Renal allograft (rat)	[76]
		Protein (i.v.)	Transplant vasculopathy (rat)	[77]
(II)	35K (VAC-L)	Protein Fc-fusion (i.d.)	Skin inflammation (guinea pig)	[54]
	vCCI(CPV)	Protein Fc-fusion (i.n.)	Airway inflammation (mouse)	[78]
	35K (VAC-L)	Adenovirus vector (i.p.)	Peritoneal inflammation (mouse)	[79]
	35K (VAC-L)	Adenovirus vector (i.p.)	Atherosclerosis (mouse)	[80]
	M-T1 (MYX)	Protein (i.v.)	Transplant vasculopathy (rat)	[77]
(III)	M3 (γ68HV)	Transgene	Pancreatic inflammation (mouse)	[81]
		Transgene	Vasculopathy (mouse)	[118]
		Protein (i.v.)	Transplant vasculopathy (rat)	[77]

viral genes are likely the consequence of theft of host immune genes, presumably by recombination with reverse-transcribed host cDNA from ancestrally infected host organisms. After a host-derived immunomodulator has been acquired by a given virus, however, subsequent evolutionary pressures can result in alterations of biologic functions of the captured modulator that are specifically advantageous to the virus [43–45]. Virus-encoded chemokines and chemokine receptors would fall into this category of pirated host immune regulators whose biologic functions have been shaped by selection pressures within virus-infected hosts [17–20].

An alternative, and more enigmatic, class of viral immunomodulators exhibits no obvious sequence relationship to any known host molecules. These orphan viral regulators have usually been discovered empirically by the ability to bind and inhibit specific host ligands. For example, the five known structural classes of viral chemokine binding proteins were all originally discovered by physical binding and inhibition assays using host chemokines, rather than by any sequence relationship with any known host chemokine or receptors [10, 15, 17–20]. For these viral regulators, they may either represent examples of independent convergent evolution or, alternately, their true relationship to host-derived genes may become revealed only as more genomic information from other organisms becomes available. In fact, some of these unique viral genes might have been originally derived from ancient host species that are now extinct, and their progenitor host genes may never be accurately documented.

This review will focus on the secreted viral chemokine regulators that have been independently expressed and utilized to treat disorders in animal models of inflammatory diseases. These secreted immunomodulators can be subdivided into virokines (ligand-like) or viroceptors (receptor-like) but it should be noted that this

Table 2 - Viral chemokine mimics tested in animal disease models

	Viral protein	Delivery mode	Animal model	Refs
(I)	vMIP-II (HHV8)	Protein (i.v.)	Glomerulonephritis (rat)	[111]
	vMIP-II (HHV8)	Plasmid (g.t.)	Cardiac allograft (mouse)	[102]
	vMIP-II (HHV8)	Protein (i.v.-op)	Spin cord injury (rat)	[110]
	vMIP-II (HHV8)	Protein (i.c.v.)	Cerebral ischemia (mouse)	[109, 108]
	vMIP-II (HHV8)	Protein (i.v.)	$CD8^+$ T-cell-dep. DTH (mouse)	[112]
(II)	MC148 (MCV)	Plasmid (g.t.)	Cardiac allograft (mouse)	[102]

thematic distinction is rather arbitrary because many were identified operationally as binding proteins or inhibitors, and operate by still-undefined mechanisms [40, 41, 46, 47]. In any event, only a small fraction of the currently known immmunoregulators from viruses have ever been tested as anti-inflammatory or anti-immune reagents in animal models [28], and the chemokine-targeted members of this group are considered in greater detail in the following sections.

Viral chemokine binding proteins

Virus-encoded chemokine inhibitors generally function as either cell surface receptor mimics, ligand mimics or as secreted chemokine binding proteins that scavenge chemokines away from host receptors at the surface of immune cells (Fig. 1). In the case of viral chemokine binding proteins (CBPs), five unrelated protein classes of such inhibitors (termed types I to V) have been reported to date, as defined by physical chemokine binding and inhibition assays [10, 15, 17–20]. Each of these five classes of CBP represent a distinctly unique protein family and the crystal structures of the two members so far reported (type II and III) reveal domain folds unrelated to any known host immune regulator [48–50].

The type I CBP is exemplified by the M-T7 protein from myxoma virus. M-T7 is a poxvirus viroceptor originally identified as a secreted 37 kDa inhibitor specific for rabbit interferon-gamma but was subsequently shown to bind with low affinity to the glycosaminoglycan (GAG) binding domain (C-terminus) of a broad spectrum of C/CC/CXC-chemokines and to inhibit leukocyte trafficking in virus-infected tissues [51–53]. Type II CBPs, also denoted as vCCIs (viral CC-chemokine inhibitors), have been isolated from a variety of poxviruses (e.g., myxoma, certain vaccinia strains, rabbitpox, and cowpox) and shown to specifically bind with high affinity and inhibit a broad spectrum of CC-chemokines [54–58]. Type III CPB is also represented by a single member, namely, the M3 protein of gamma-68 herpesvirus, which binds and inhibits members of all four classes of chemokines and both

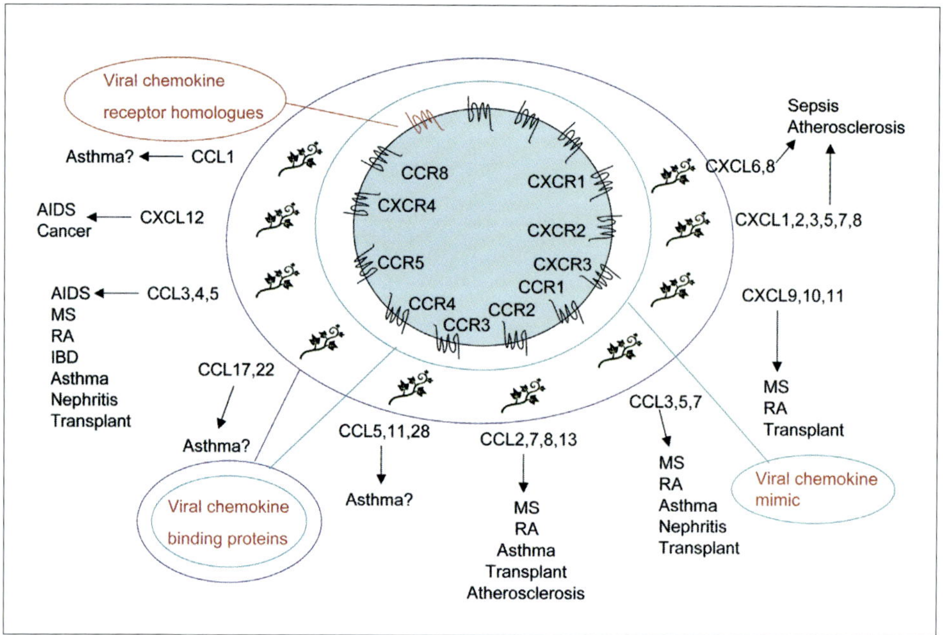

Figure 2

Chemokines, chemokine receptors and related diseases

Abbreviations: MS, multiple sclerosis; RA, rheumatoid arthritis; IBD, inflammatory bowel disease. Adapted from: Proudfoot AEI (2002) Chemokine receptors: multifaceted therapeutic targets. Nat Rev Immunol *2(2): 106–115*

occludes chemokine interactions with host receptors and the GAG elements responsible for chemokine gradients [49, 59–65]. The Type IV CBPs were recently reported in several alpha-herpesviruses, in that certain isoforms of glycoprotein G were shown to possess the ability to bind and inhibit a wide spectrum of C/CC/CXC-chemokines [66]. A single type V CBP, pUL21.5 of human cytomegalovirus, exhibits unusual specificity only for RANTES [67]. Overall, extensive work is currently underway in the academic and pharmaceutical worlds in the development of novel chemokine-modulatory drugs, and the known viral CBPs represent a potent repository of reagents with which to manipulate chemokine functions and leukocyte trafficking [68–74].

The three classes of viral CBPs that have been experimentally tested to date in animal models (i.e., CBP I–III) have each demonstrated clearly the elegant sophistication that viruses have evolved to thwart the chemokine circuitry (Tab. 1). *In vivo* studies with purified CBPs I–III have consistently demonstrated effective inhibition of inflammatory disorders in a range of animal disease models. The Type I CBP, M-

T7, was shown to block invasion of macrophages and T lymphocytes following vascular injury in rat and rabbit models [75–77]. Vascular balloon angioplasty injury and aortic allograft transplant models were both utilized to initiate a marked arterial inflammatory response, which is particularly aggressive following aortic transplantation. Also, with these models inflammation in the vascular wall is the therapeutic target. When the arterial wall is studied as a target for anti-inflammatory chemokine response modifying agents, we are in fact studying the initial site of entry for inflammatory cells heading to injured organs or tissues (i.e., the initiation point). The inflammatory response in the arterial wall thus allows one to assess the effects of viral anti-inflammatory proteins at a site where many innate immune system responses originate. Infusion of purified M-T7 protein resulted in the inhibition of early mononuclear cell invasion post-injury and was associated with long term reductions in atherosclerotic plaque growth (vasculopathy) following either transplant or balloon angioplasty injury or stent implant [75–77]. The lack of species specificity of M-T7 in the various species of animal models tested suggests that the inhibition of cell invasion and plaque growth was in fact the consequence of targeting the host chemokines rather than inteferon-γ, whose inhibition by M-T7 is restricted to the rabbit species [75–77]. Furthermore, M-T7 protein suppressed the vascular pathology associated with inflammatory disease models even when given transiently at very low dosages (pg-ng/kg body weight). Bedard et al. similarly demonstrated that intravenous treatment with M-T7 protein, given daily at doses up to 80 ng/kg for only the first 10 days post transplant, markedly reduced vasculopathy and organ scarring in rat renal transplants as long as five months after surgery [76].

The viral CBP type II, M-T1 from myxoma virus, which shares close sequence similarity to vCCI/35K from vaccinia, has also been tested in rat and mouse aortic allograft models. Using the rat model, M-T1 protein, when (given intravenously as a single protein bolus administered immediately following vascular transplant, mediated blockade of early mononuclear cell invasion and also inhibited the development of chronic transplant vasculopathy [77].

Using the related chemokine binding protein vCCI/35K, Dabbagh et al. demonstrated that infusion of vCCI/35K as an Fc-fusion protein significantly reduced airway inflammation in a mouse model [78]. vCCI/35k also initiated eosinophil influx associated with eotaxin-mediated inflammation in the guinea pigs skin model [54, 77]. When expressed from an adenovirus vector that was delivered by intra-peritoneal injection, vCCI/35K downregulated inflammatory cell recruitment induced by biogel in peritoneal exudates in mice and also reduced plaque development in ApoE-knockout mice [79, 80]. M3, a class III CBP also displayed potent therapeutic activity, blocking aortic allograft vasculopathy [77] and pancreatic inflammation [81]. Significantly, when M3 expression was conditionally upregulated in a mouse model of femoral arterial injury, a significant reduction in intimal hyperplasia was also detected in this model [118].

Of interest, in the angioplasty injury models, the administration of a glycosaminoglycan frequently utilized for clinical clotting disorders (heparan sulfate) was capable of blocking some of the anti-inflammatory activity of M-T7, presumably by interfering with M-T7 binding to the GAG binding domain of chemokines. This further supports a chemokine based mechanism of inhibitory action for M-T7 [75]. In the aortic transplant model, the infusion of the CC chemokines MCP-1 or MIP-1α selectively blocked M-T1 and M-T7 mediated inhibition of arterial monocyte invasion, respectively, after transplant [77]. Combined treatment with M-T1 and M-T7 at higher doses did not result in greatly enhanced anti-inflammatory and anti-atherogenic activity again indicating overlapping targets (specifically CC chemokines) and activities [77].

The analysis of these three diverse classes of viral CBPs reaffirms the importance and impact of the chemokine system on both early inflammatory responses to trauma and as well as long-term disease development. Whether administered as purified proteins [54, 75–78], expressed through adenoviral vectors [79, 80], or produced endogenously in transgenic mice [81, 118], CBPs could consistently induce profound inhibition of inflammatory responses to a wide spectrum of inducers. CBPs also provide powerful tools to deconstruct the critical roles that chemokines play during inflammatory responses, but note that the actual mechanisms through which these viral CBPs functionally block chemokine responses when given in such relatively low doses for very restricted time frames still is not fully understood. For example, the CBP type I, M-T7, inhibits inflammatory influx effectively *in vivo* at very low protein dosages [75–77] whereas this protein binds the GAG binding domain of chemokines with only low affinity *in vitro* [51]. In contrast, M-T1, which binds the N-terminus, blocks leukocyte migration in Boyden chamber assays in response to soluble CC chemokine gradient stimulation, while M-T7 is ineffective as an inhibitor in this model. M-T7, on the other hand, is quite effective at blocking mononuclear cell invasion into mouse ascites in response to intraperitoneal injection of the CC chemokine, MCP-1 (A. Lucas, unpublished results), once again supporting an anti-inflammatory function for M-T7 that is mediated through disruption of chemokine gradient formation. The low affinity GAG-binding domains of many chemokines are critical for gradient stabilization and ligand presentation to the invading leukocytes but it is nevertheless quite surprising that this protein would be so effective at such low concentrations *in vivo*. Instead, one conclusion would be that the GAG/chemokine interaction is a tractable target for pharmacologic intervention [38, 82–84]. This opinion is, in fact, rapidly being confirmed by studies with inhibitors that block chemokine GAG binding actions [82].

Less work has been performed to date assessing the potential for use of viral chemokine receptors that bind chemokines in modifying pathology [85]. Expression of the HHV8 Kaposi's sarcoma associated herpesvirus ORF74 in transgenic mice did not reduce inflammatory disease, but in fact, resulted in angioproliferative lesions and enhanced tumorigenesis in multiple organs that in fact resembled Kaposi's sarcomas [86]. The cytomegalovirus (CMV) encoded G protein coupled chemokine receptor, US28 induces smooth muscle cellular migration which has the potential to accelerate atheroma development [87–89]. Thus, these viral chemokine receptors do have the potential to exacerbate or initiate vascular disease states and to date have not yet been shown to ameliorate inflammatory or immune responses in animal models of disease.

Viral chemokine mimics

In the case of viral chemokine mimics, the two examples that have been tested to date in animal models of inflammation are MC148 of *Molluscum contagiosum* virus (MCV) and vMIP-II of human herpesvirus-8/Kaposi's Sarcoma Herpes Virus (HHV-8/KSHV) (Tab. 2). MC148 of MCV exhibits significant specificity for human CCR8 and antagonizes the lone host chemokine ligand that signals via this receptor (I-309), whereas vMIP-II is both an agonist for CCR3 and a promiscuous antagonist for at least ten human CC- and CXC-chemokine receptors [90–101]. Unlike vMIP-II, MC148 does not bind any known murine chemokine receptors, and would not be predicted to be anti-inflammatory in mouse models. Nevertheless, the available data indicates that both MC148 and vMIP–II can each prolong cardiac allograft survival in mice [102]. vMIP-II also can block Th1-polarized T-lymphocytes while stimulating Th2-responses, thereby downregulating cell-mediated immune responses [100, 103]. At present, it has not been elucidated how MC148 inhibits inflammation in the murine system but it is possible that it also targets inflammatory pathways independent of the chemokine system. Alternatively, there could be still-to-be identified chemokine receptors on primary cells that are in fact antognized by MC148 [104]. The structure of MC148 protein has not yet been reported, but vMIP-II is reported to be a monomeric protein with many chemokine-like canonical folds [105].

In the brain, chemokines are expressed at elevated concentrations after mechanical trauma or chronic neuropathic disorders such as Alzheimer's disease and multiple sclerosis [106–108]. Takami et al. have shown that intracerebroventricular injections of purified vMIP-II protein, which can antagonize macrophage inflammatory protein-1α (MIP-1α, or CCL3), reduced the size of the cerebral infarct at 48 h after middle cerebral arterial occlusion whereas, injection of MIP-1α increased infarct size in mice [109]. Ghirnikar et al. similarly found that infusion of vMIP-II protein for 7 days via osmotic minipump brain infusion following spinal cord contusion in rats resulted in a decrease in the number of infiltrating neutrophils (day 1 post injury), macrophages (days 3–7 post injury), and microglia (days 3–7 post injury) [110]. The reduction in inflammatory cell invasion was associated with lower levels of neuronal loss and increased expression of Bcl-2, an endogenous apoptosis inhibitor [110]. In a rat model of glomerulonephritis, intravenous infusion of vMIP-II protein downregulated CC and CX3C chemokine expression, reduced macrophage and T lymphocyte invasion, and resulted in less crescentic glomeruli and proteinuria (protein loss in the urine indicative of kidney damage) [111]. Inflammatory exudates, that are thought to generate some of the CD8+T cell mediated immunopathology associated with lymphocytic choriomeningitis virus infections, were also reduced following vMIP-II treatment in mice [112]. In a mouse cardiac allograft transplant model, gene transfer of vMIP-II or MC148 reduced cytotoxic T lymphocyte infiltrates and alloantibody production with associated prolonged graft survival (survival for 21 days with vMIP-II *versus* 13 days for control) [102]. Injection of purified vIL-10 protein together with vMIP-II further enhanced graft survival, suggesting these viral immunomodulating cytokines inhibited inflammatory responses through synergistic pathways [102]. Recently, it has been shown that vMIP-II possesses unique properties distinct from the cellular CCR8 ligands (I-309 and TCA-3) in terms of mucosal Th2 responses, IL-10 regulation and host co-stimulator expression [113].

Future prospects

As long-term inquisitors of the mammalian immune system, viruses have developed an extraordinary range of virally mediated immunomodulatory agents. Through the unraveling of viral anti-immune strategies targeted against the host chemokine networks, a new class of therapeutic agents has been revealed based upon virus-engineered chemokine-modulatory proteins. Viruses were the first organisms for which complete genome sequences were deduced, beginning a quarter of a century ago, and the science of "virogenomics" has been expanding rapidly ever since [114–116]. The repertoire of novel viral gene products that are devoted to host modulation has also been proliferating at an astounding rate, and there are reasons to suspect that we have uncovered only the tip of the virus iceberg. For example, the discipline of virology has largely focused on viruses that cause overt pathogenesis but the viral ecosphere is populated largely by apathogenic members that still remain to be discovered. Indeed, there are proposals to fully define the complete human "virome", or the summated sequences of all viruses that are present in the human population [117], and such genomic mining will likely uncover an even greater armamentarium of viral immuno-regulators.

We project that immunomodulatory viral proteins targeted against chemokines and their receptors will establish a new platform for treatment of inflammation based disorders. As more is learned about how these virus-derived drug candidates behave as pharmacological reagents, particularly in human clinical trials, we will be in better position to evaluate which human diseases have the potential to be effectively treated with this novel class of biopharmaceuticals.

Acknowledgements
We thank Doris Hall for assistance with the manuscript preparation. GM and AL are co-founders of VIRON Therapeutics, which is developing viral proteins as anti-inflammatory therapeutics. GM holds a Canada Research Chair in Molecular Virology. The authors' labs are funded by the CIHR, NCIC and the Heart and Stroke Foundation.

References

1 Alcami A, Koszinowski UH (2000) Viral mechanisms of immune evasion. *Immunol Today* 21: 447–455
2 Tortorella D, Gewurz BE, Furman MH, Schust DJ, Ploegh HL (2000) Viral subversion of the immune system. *Ann Rev Immunol* 18: 861–926
3 Agrawal N, Korkaya H, Jameel S (2000) How viruses evade host responses [Review]. *Curr Sci* 79: 711–724
4 Rosenkilde MM, Walkhoer M, Luttichau HR, Schwartz TW (2001) Virally encoded 7TM receptors. *Oncogene* 20: 1582–1593
5 Xu X-N, Screaton GR, McMichael AJ (2001) Virus Infections: Escape resistance, and counterattack. *Immunity* 15: 867–870
6 Lorenzo ME, Ploegh HL, Tirabassi RS (2001) Viral immune evasion strategies and the underlying cell biology. *Immunology* 13: 1–9
7 Gewurz BE, Gaudet R, Tortorella D, Wang EW, Ploegh HL (2001) Virus subversion of immunity: A structural perspective. *Curr Op Immunol* 13: 442–450
8 Vossen MTM, Westerhout EM, Soderberg-Naucler C, Wiertz EJHJ (2002) Viral immune evasion: A masterpiece of evolution. *Immunogenetics* 54: 527–542
9 Petersen JL, Morris CR, Solheim JC (2003) Virus evasion of MHC Class I molecule presentation. *J Immunol* 171: 4473–4478
10 Alcami A (2003) Viral mimicry of cytokines, chemokines and their receptors. *Nature Rev Immunol* 3: 36–50
11 McFadden G, Murphy PM (2000) Host-related immunomodulators encoded by poxviruses and herpesviruses. *Curr Op Microbiol* 3: 371–378
12 Smith GL (2000) Secreted poxvirus proteins that interact with the immune system In: RS Fujinami (ed): *Effects of Microbes on the Immune System*. Lippincott Williams and Wilkins, Philadelphia, 491–507
13 Moss B, Shisler JL (2001) Immunology 101 at poxvirus U: Immune evasion genes. *Semin Immunol* 13: 59–66
14 Johnston JB, McFadden G (2003) Poxvirus immunomodulatory strategies: Current perspectives. *J Virol* 77: 6093–6100
15 Seet BT, Johnston JB, Brunetti CR, Barrett JW, Everett H, Cameron C, Sypula J, Nazarian SH, Lucas A, McFadden G (2003) Poxviruses and immune evasion. *Ann Rev Immunol* 21: 377–423

16 Shchelkunov SN (2003) Immunomodulatory proteins of orthopoxviruses. *Mol Biol* 37: 37–48
17 Lalani AS, Barrett J, McFadden G (2000) Modulating chemokines: More lessons from viruses. *Immunol Today* 21: 100–106
18 Murphy PM (2001) Viral exploitation and subversion of the immune system through chemokine mimicry. *Nature* 2: 116–122
19 Seet BT, McFadden G (2002) Viral chemokine binding proteins. *J Leukocyte Biol* 72: 24–34
20 Holst PJ, Rosenkilde MM (2003) Microbiological exploitation of the chemokine system. *Microb Infect* 5: 179–187
21 Saederup N, Mocarski J (2002) Fatal attraction: Cytomegalovirus-encoded chemokine homologs. *Curr Top Microbiol Immunol* 269: 235–256
22 Kotwal GJ (2000) Poxviral mimicry of complement and chemokine system components: What's the end game? *Immunol Today* 21: 242–248
23 Liston A, McColl S (2003) Subversion of the chemokine world by microbial pathogens. *BioEssays* 25: 478–488
24 Dairaghi DJ, Greaves DR, Schall TJ (1998) Abduction of chemokine elements by herpesviruses. *Semin Virol* 8: 377–385
25 Chensue SW (2001) Molecular machinations: Chemokine signals in host-pathogen interactions. *Clin Microbiol Rev* 14: 821–835
26 Lalani AS, McFadden G (1999) Evasion and exploitation of chemokines by viruses. *Cytokine Growth Factor Rev* 10: 219–233
27 Sodhi A, Mantaner S, Gutkind JS (2004) Viral hijacking of G-protein coupled-receptor signalling networks. *Nature Rev/Mol Cell Biol* 5: 998–1012
28 Lucas A, McFadden G (2004) Secreted immunomodulatory viral proteins as novel biotherapeutics. *J Immunol* 173: 4765–4774
29 Murdoch C, Finn A (2000) Chemokine receptors and their role in inflammation and infectious disease. *Blood* 95: 3032–3043
30 Zlotnik A, Yoshie O (2000) Chemokines: A new classification system and their role in immunity. *Immunity* 12: 121–127
31 Rot A, Von Andrian UH (2004) Chemokines in innate and adaptive host defense: Basic chemokinese grammar for immune cells. *Ann Rev Immunol* 22: 891–928
32 Moser B, Wolf M, Walz A, Loetscher P (2004) Chemokines: Multiple levels of leukocyte migration control. *Trends Immunol* 25: 75–84
33 Mackay CR (2001) Chemokines: Immunology's high impact factor. *Nature* 2: 95–101
34 Fernandez EJ, Lolis E (2001) Structure, function, and inhibition of chemokines. *Ann Rev Pharmacol Tox* 42: 469–499
35 Proudfoot AEI (2002) Chemokine receptors: Multifaceted therapeutic targets. *Nature Rev Immunol* 2: 106–115
36 Ali S, Palmer ACV, Banerjee B, Fritchley SJ, Kirby JA (2000) Examination of the function of RANTES, MIP-1a, and MIP-1b following interaction with heparin-like glycosaminoglycans. *J Biol Chem* 275: 11721–11727

37 Campbell DJ, Kim CH, Butcher EC (2003) Chemokines in the systemic organization of immunity. *Immunol Rev* 195: 58–71
38 Johnson Z, Power CA, Weiss C, Rintelen F, Ji H, Rickle T, Camps M, Wells TNC, Schwarz MK, Proudfoot AEI et al (2004) Chemokine inhibition – Why, when, where, which and how? *Biochem Soc Trans* 32: 366–377
39 Charo IF, Taubman MB (2004) Chemokines in the pathogenesis of vascular disease. *Circ Res* 95: 858–866
40 Barry M, McFadden G (1997) Virokines and viroceptors In: JS Friedland (ed): *Cytokines in Health and Disease*. Marcel Dekker Inc, New York, 251–261
41 Barry M, McFadden G (1997) Virus encoded cytokines and cytokine receptors. *Parasitology* 115: S89–S100
42 Alcamí A, Koszinowski UH (2000) Viral mechanisms of immune evasion. *Immunol Today* 21: 447–455
43 Upton C, Slack S, Hunter AL, Ehlers A, Roper RL (2003) Poxvirus orthologous clusters: Toward defining the minimum essential poxvirus genome. *J Virol* 77: 7590–7600
44 McLysaght A, Baldi PF, Gaut BS (2003) Extensive gene gain associated with adaptive evoluton of poxviruses. *Proc Nat Acad Sci USA* 100: 15655–15660
45 Gubser C, Hue S, Kellam P, Smith GL (2004) Poxvirus genomes: A phylogenetic analysis. *J Gen Virol* 85: 105–117
46 Smith SA, Kotwal GJ (2001) Virokines: Novel immunomodulatory agents. *Expert Op Biol Ther* 1: 343–357
47 McFadden G, Lalani A, Everett H, Nash P, Xu X (1998) Virus encoded-receptors for cytokines and chemokines. *Semin Cell Develop Biol* 9: 359–368
48 Carfi A, Smith CA, Smolak PJ, McGrew J, Wiley DC (1999) Structure of a soluble secreted chemokine inhibitor vCCI (p35) from cowpox virus. *Proc Nat Acad Sci USA* 96: 12379–12383
49 Alexander JM, Nelson CA, Van Berkel V, Lau EK, Studts JM, Brett TJ, Speck SH, Handel TM, Virgin HW, Fremont DH (2002) Structural basis of chemokine sequestration by a herpesvirus decoy receptor. *Cell* 111: 343–356
50 Alcami A (2003) Structural basis of the herpesvirus M3-chemokine interaction. *Trends Microbiol* 11: 191–192
51 Lalani AS, Graham K, Mossman K, Rajarathnam K, Clark-Lewis I, Kelvin D, McFadden G (1997) The purified myxoma virus gamma interferon receptor homolog, M-T7, interacts with the heparin binding domains of chemokines. *J Virol* 71: 4356–4363
52 Mossman K, Nation P, Macen J, Garbutt M, Lucas A, McFadden G (1996) Myxoma virus M-T7, a secreted homolog of the interferon-g receptor, is a critical virulence factor for the development of myxomatosis in European rabbits. *Virol* 215: 17–30
53 Upton C, Mossman K, McFadden G (1992) Encoding of a homolog of the IFN-gamma receptor by myxoma virus. *Science* 258: 1369–1372
54 Alcamí A, Symons JA, Collins PD, Williams TJ, Smith GL (1998) Blockade of chemokine activity by a soluble chemokine binding protein from vaccinia virus. *J Immunol* 160: 624–633

55 Smith CA, Smith TD, Smolak PJ, Friend D, Hagen H, Gerhart M, Park L, Pickup DJ, Torrance D, Mohler K et al (1997) Poxvirus genomes encode a secreted soluble protein that preferentially inhibits b chemokine activity yet lacks sequence homology to known chemokine receptors. *Virol* 236: 316–327
56 Graham KA, Lalani AS, Macen JL, Ness TL, Barry M, Liu L-Y, Lucas A, Clark-Lewis I, Moyer RW, McFadden G (1997) The T1/35kDa family of poxvirus secreted proteins bind chemokines and modulate leukocyte influx into virus infected tissues. *Virol* 229: 12–24
57 Lalani AS, Ness TL, Singh R, Harrison JK, Seet BT, Kelvin DJ, McFadden G, Moyer RW (1998) Functional comparisons among members of the poxvirus T1/35kDa family of soluble CC-chemokine inhibitor glycoproteins. *Virology* 250: 173–184
58 Reading PC, Symons JA, Smith GL (2003) A soluble chemokine-binding protein from vaccinia virus reduces virus virulence and the inflammatory response to infection. *J Immunol* 170: 1435–1442
59 Van Berkel V, Barrett J, Tiffany HL, Fremont DH, Murphy PM, McFadden G, Speck SH, Virgin HW (2000) Identification of a gammaherpesvirus selective chemokine binding protein that inhibits chemokine action. *J Virol* 74: 6741–6747
60 Parry BC, Simas JP, Smith VP, Stewart CA, Minson AC, Efstathiou S, Alcamí A (2000) A broad spectrum secreted chemokine binding protein encoded by a herpesvirus. *J Exp Med* 191: 573–578
61 Alcami A, Efstathiou S (2000) Soluble chemokine binding proteins are also encoded by herpesviruses. *Immunol Today* 21: 526–527
62 Webb LMC, Clark-Lewis I, Alcami A (2003) The gammaherpesvirus chemokine binding protein binds to the N terminus of CXCL8. *J Virol* 77: 8588–8592
63 Webb LMC, Smith VP, Alcami A (2004) The gammaherpesvirus chemokine binding protein can inhibit the interaction of chemokines with glycosaminoglycans. *FASEB J* 18: 571–573
64 Bridgeman A, Stevenson PG, Simas JP, Efstathiou S (2001) A secreted chemokine binding protein encoded by murine gammaherpesvirus-68 is necessary for the establishment of a normal latend load. *J Exp Med* 194: 301–312
65 Van Berkel V, Levine B, Kapadia SB, Goldman JE, Speck SH, Virgin HWT (2002) Critical role for a high-affinity chemokine-binding protein in g-herpesvirus-induced lethal meningitis. J Clin Invest 109: 905–914
66 Bryant NA, Davis-Poynter N, Vanderplasschen A, Alcami A (2003) Glycoprotein G isoforms from some alphaherpesviruses function as broad-spectrum chemokine binding proteins. *EMBO J* 22: 833–846
67 Wang D, Bresnahan W, Shenk T (2004) Human cytomegalovirus encodes a highly specific RANTES decoy receptor. *Proc Nat Acad Sci USA* 101: 16642–16647
68 Lindow M, Luttichau HR, Schwartz TW (2003) Viral leads for chemokine-modulatory drugs. Trends Pharmacol Sci 24: 126–130
69 Proudfoot AEI, Power CA, Rommel C, Wells TNC (2003) Strategies for chemokine antagonists as therapeutics. *Semin Immunol* 15: 57–65

70 Schwarz MK, Wells TN (2002) New therapeutics that modulate chemokine networks. *Nature Rev Drug Dis* 1: 347–358

71 Houshmand P, Zlotnik A (2003) Therapeutic applications in the chemokine superfamily. *Curr Op Chem Biol* 7: 457–460

72 Grainer DJ, Reckless J (2003) Broad-spectrum chemokine inhibitors (BSCIs) and their anti-inflammatory effects *in vivo*. *Biochem Pharmacol* 65: 1027–1034

73 Onuffer JJ, Horuk R (2002) Chemokines, chemokine receptors and small-molecule antagonists: Recent developments. *Trends Pharmacol Sci* 23: 459–467

74 Rajarathnam K (2002) Designing decoys for chemokine-chemokine receptor interaction. *Curr Phama Design* 8: 2159–2169

75 Liu LY, Lalani A, Dai E, Seet B, Macauley C, Singh R, Fan L, McFadden G, Lucas A (2000) The viral anti-inflammatory chemokine-binding protein M-T7 reduces intimal hyperplasia after vascular injury. *J Clin Invest* 105: 1613–1621

76 Bedard ELR, Kim P, Jiang J, Parry N, Liu L, Wang H, Garcia B, Li X, McFadden G, Lucas A et al (2003) Chemokine-binding viral protein M-T7 prevents chronic rejection in rat renal allografts. *Transplantation* 76: 249–252

77 Liu L, Dai E, Miller L, Seet B, Lalani A, Macauley C, Li X, Virgin HW, Bunce C, Turner P et al (2004) Viral chemokine-binding proteins inhibit inflammatory responses and aortic allograft transplant vasculopathy in rat models. *Transplantation* 77: 1652–1660

78 Dabbagh K, Xiao Y, Smith C, Stepick-Biek P, Kim SG, Lamm WJ, Liggitt DH, Lewis DB (2000) Local blockade of allergic airway hyperreactivity and inflammation by the poxvirus-derived pan-CC-chemokine inhibitor vCCI. *J Immunol* 165: 3418–3422

79 Bursill CA, Cai S, Channon KM, Greaves DR (2003) Adenoviral-mediated delivery of a viral chemokine binding protein blocks CC-chemokine activity *in vitro* and *in vivo*. *Immunol* 207: 187–196

80 Bursill CA, Choudbury RP, Ali Z, Greaves DR, Channon KM (2004) Broad-spectrum CC-chemokine blockade by gene transfer inhibits macrophage recruitment and atherosclerotic plaque formation in apolipopritein E-knockout mice. *Circulation* 110: 2460–2466

81 Jensen KK, Chen S-C, Hipkin RW, Wiekowski MT, Schwarz MA, Chou C-C, Simas JP, Alcami A, Lira SA (2003) Disruption of CCL21-induced chemotaxis *in vitro* and *in vivo* by M3, a chemokine-binding protein encoded by murine gammaherpesvirus 68. *J Virol* 77: 624–630

82 Proudfoot AEI, Handel TM, Johnson Z, Lau EK, LiWang P, Clark-Lewis I, Borlat F, Wells TNC, Kosco-Vilbois MH (2003) Glycosaminoglycan binding and oligomerization are essential for the *in vivo* activity of certain chemokines. *Proc Nat Acad Sci USA* 100: 1885–1890

83 Lau EK, Paavola CD, Johnson Z, Gaudry J-P, Geretti E, Borlaat F, Kungl AJ, Proudfoot AE, Handel TM (2004) Identification of glycosaminoglycan binding site of the CC chemokine, MCP-1; implications for structure and function *in vivo*. *J Biol Chem* 279: 22294–22305

84 Johnson Z, Kosco-Vilbois MH, Herren S, Cirillo R, Muzio V, Zaratin P, Carbonatto M,

Mack M, Smailbegovic A, Rose M et al (2004) Interference with heparin binding and oligomerization creates a novel anti-inflammatory strategy targeting the chemokine system. *J Immunol* 173: 5776–5785
85 Smit MJ, Vink C, Verzijl D, Casarosa P, Bruggeman CA, Leurs R (2003) Virally encoded G protein-coupled receptors: Targets for potentially innovative anti-viral drug development. *Curr Drug Targets* 4: 431–441
86 Yang BT, Chen SC, Leach MW, Manfra D, Homey B, Wiekowski M, Sullivan L, Jenh CH, Narula SK, Chensue SW et al (2000) Transgenic expression of the chemokine receptor encoded by human herpesvirus 8 induces an angioproliferative disease resembling Kaposi's Sarcoma. *J Exp Med* 191: 445–454
87 Streblow DN, Orloff SL, Nelson JA (2001) The HCMV chemokine receptor US28 is a potential target in vascular disease. *Curr Drug Target Infect Dis* 1: 151–158
88 Streblow DN, Vomaske J, Smith P, Melnychuk R, Hall L, Pancheva D, Smit M, Casarosa P, Schlaepfer DD, Nelson JA (2003) Human cytomegalovirus chemokine receptor US28-induced dsmooth muscle cell migration is mediated by focal adhesion kinase and Src. *J Biol Chem* 278: 50456–50465
89 Melnychuk RM, Streblow DN, Smith PP, Hirsch AJ, Pancheva D, Nelson JA (2004) Human cytomegalovirus-encoded G protein-coupled receptor US28 mediates smooth muscle cell migration through Galpha12. *J Virol* 78: 8283–8391
90 Luttichau BH, Stine J, Boesen TP, Johnsen AH, Chanry D, Gerstoft J, Schwartz TW (2000) A highly selective CC chemokine receptor (CCR)8 antagonist encoded by the poxvirus Molluscum Contagiosum. *J Exp Med* 191: 171–180
91 Luttichau HR, Gerstoft J, Schwartz TW (2001) MC148 encoded by human molluscum contagiosum poxvirus is an antagonist for human but not murine CCR8. *J Leukocyte Biol* 70: 277–282
92 Laing KJ, Secombes CJ (2004) Chemokines. *Dev Compar Immunol* 28: 443–460
93 Haque NS, Fallon JT, Pan JJ, Taubman MB, Harpel PC (2004) Chemokine receptor-8 (CCR8) mediates human vascular smooth muscle cell chemotaxis and metalloproteinase-2 secretion. *Blood* 103: 1296–1304
94 Littichau HR, Lewis IC, Gerstoft J, Schwartz TW (2001) The herpesvirus 8-encoded chemokine vMIP-II, but not the poxvirus-encoded chemokine MC148, inhibits the CCR10 receptor. *Eur J Immunol* 31: 1217–1220
95 Spinetti G, Bernardini G, Camarda G, Mangoni A, Santoni A, Capogrossi MC, Napolitano M (2003) The chemokine receptor CCR8 mediates rescue from dexamethasone-induced apoptosis via an ERK-dependent pathway. *J Leukocyte Biol* 73: 201–207
96 Zhou N, Luo Z, Luo J, Hall JW, Huang Z (2000) A novel peptide antagonist of CXCR4 derived from the N-terminus of viral chemokine vMIP-II. *Biochem* 39: 3782–3787
97 Shan LX, Qiao XD, Oldham E, Catron D, Kaminski H, Lundell D, Zlotnik A, Gustafson E, Hedrick JA (2000) Identification of viral macrophage inflammatory protein (vMIP)-II as a ligand for GPR5/XCR1. *Biochem Biophys Res Comm* 268: 938–941
98 Kledal TN, Rosenkilde MM, Coulin F, Simmons G, Johnsen AH, Alouani S, Power CA,

Luttchau HR, Gerstoft J, Clapham PR et al (1997) A broad-spectrum chemokine antagonist encoded by Kaposi's sarcoma-associated herpesvirus. *Science* 277: 1656–1659

99 Boshoff C, Endo Y, Collins PD, Takeuchi Y, Reeves JD, Schweickart VL, Siani MA, Sasaki T, Williams TJ, Gray PW et al (1997) Angiogenic and HIV-inhibitory functions of KSHV-encoded chemokines. *Science* 278: 290–294

100 Sozzani S, Luini W, Bianchi G, Allavena P, Wells TN, Napolitano M, Bernardini G, Vecchi A, D'Ambrosio D, Mazzeo D et al (1998) The viral chemokine macrophage inflammatory protein-II is a selective Th2 chemoattractant. *Blood* 92: 4036–4039

101 Nakano K, Isegawa Y, Zou P, Tadagaki K, Inagi R, Yamanishi K (2003) Kaposi's sarcoma-associated herpesvirus (KSHV)-encoded vMIP-I and vMIP-II induce signal transduction and chemotaxis in monocytic cells. *Arch Virol* 145: 871–890

102 deBruyne LA, Li K, Bishop DK, Bromberg JS (2000) Gene transfer of virally encoded chemokine antagonists vMIP-11 and MC148 prolongs cardiac allograft survival and inhibits donor-specific immunity. *Gene Ther* 7: 575–582

103 Weber KSC, Grone HJ, Rocken M, Klier C, Gu SH, Wank R, Proudfoot AEI, Nelson PJ, Weber C (2001) Selective recruitment of Th2-type cells and evasion from a cytotoxic immune response mediated by viral macrophage inhibitory protein-II. *Eur J Immunol* 31: 2458–2466

104 Damon I, Murphy PM, Moss B (1998) Broad spectrum chemokine antagonistic activity of a human poxvirus chemokine homolog. *Proc Nat Acad Sci USA* 95: 6403–6407

105 Liwang AC, Wang Z-X, Sun Y, Peiper SC, Liwang PJ (1999) The solution structure of the anti-HIV chemokine vMIP-II. *Protein Sci* 8: 2270–2280

106 Gerard C, Rollins BJ (2001) Chemokines and disease. Nature Immunol 2: 108–115

107 Tran PB, Miller RJ (2003) Chemokine receptors: Signposts to brain development and disease. *Nature Rev Neurosci* 4: 444–455

108 Minami M, Satoh M (2003) Chemokines and their receptors in the brain: Pathophysiological roles in ischemic brain injury. *Life Sci* 74: 321–327

109 Takami S, Minami M, Nagata I, Namura S, Satoh M (2001) Chemokine receptor antagonist peptide, viral MIP-II, protects the brain against focal cerebral ischemia in mice. *J Cereb Blood Flow Metab* 21: 1430–1435

110 Ghirnikar RS, Lee YL, Eng LF (2000) Chemokine antagonist infusion attenuates cellular infiltration following spinal cord contusion injury in rat. *J Neurosci Res* 59: 63–73

111 Chen S, Bacon KB, Li L, Garcia GE, Xia Y, Lo D, Thompson DA, Siani MA, Yamamoto T, Harrison JK et al (1998) *In vivo* inhibition of CC and CX3C chemokine-induced leukocyte infiltration and attention of glomerulonephritis in Wistar-Kyoto (WKY) rats by vMIP-II. *J Exp Med* 188: 193–198

112 Lindow M, Nansen A, Bartholdy C, Stryhn A, Hansen NJV, Boesen TP, Wells TNC, Schwartz TW, Thomsen AR (2003) The virus-encoded chemokine vMIP-II inhibits virus-induced Tc1-driven inflammation. *J Virol* 77: 7393–7400

113 Singh UP, Singh S, Ravichandran P, Taub DD, Lillard JW Jr (2004) Viral macrophage-inflammatory protein-II: A viral chemokine that differentially affects adaptive mucosal immunity compared with its mammalian counterparts. *J Immunol* 173: 5509–5516

114 Fruh K, Simmen K, Luukkonen BGM, Bell YC, Ghazal P (2001) Virogenomics: A novel approach to antiviral drug discovery. *Res Focus Rev* 6: 621–627
115 DeFilippis V, Raggo C, Moses A, Fruh K (2003) Functional genomics in virology and antiviral drug discovery. *Trends Biotech* 21: 452–457
116 Kellam P (2001) Post-genomic virology: The impact of bioinformatics, microarrays and proteomics on investigating host and pathogen interactions. Rev Med Virol 11: 313–329
117 Anderson NG, Gerin JL, Anderson NL (2003) Global screening for human viral pathogens. *Emerg Infect Dis* 9: 768–773
118 Pyo R, Jensen KK, Wiekowski MT, Manfra D, Alcami A, Taubman MB, Lira SA (2004) Inhibition of intimal hyperplasia in transgenic mice conditionally expressing the chemokine binding protein M3. *Am J Pathol* 164: 2289–2297

Chemokine receptors in tissue cells and angiogenesis

Paola Romagnani, Laura Lasagni, and Sergio Romagnani

Center of Excellence for Research, Transfer and High Education DENOthe of the University of Florence, Viale Morgagni 85, Firenze 50134, Italy

Introduction

Although chemokines have been initially discovered and universally known as cytokines able to recruit leukocytes to inflamed tissues (chemotactic cytokines) and, therefore, to play an important role in the context of the immune response, subsequent studies have clearly shown that they also act on several other cell types, thus behaving as multifunctional mediators. The nature and classification of chemokines, their receptors and signalling pathways, as well as their activity of recruitment on the cells of the immune system have been discussed in other chapters of this book. Here, therefore, we will concentrate on the production of chemokines by, and on their functional activity on, tissue cells, and we will particularly focus on the essential role of chemokines on the induction and control of angiogenesis.

Chemokines in embryogenesis

Cell migration is an integral component of embryogenesis, particularly since cell position is a primary determinant of cell fate. Not surprisingly, there are complex arrays of regulators, which direct cell movement by modulating adhesion, attraction, and repulsion. Several chemokine receptors have been found to be expressed in the mouse embryo, the message encoding CXCR4 being the predominant chemokine receptor detected [1]. CXCR4- and CXCL12-deficient mice [2, 3] showed defects in the development of neuronal, cardiac, vascular, haemopoietic and craniofacial systems. Other chemokine receptor messages were also found, but all of them concordant temporally and spatially with definitive (adult-like) haematopoiesis. CX3CL1, CXCL10 and CXCL12 are certainly involved in the development of human kidney, CX3CL1 being strongly expressed during glomerulogenesis, while CXCL10 and CXCL12 in developing kidneys were more limited to focal expression [4]. More recently, CXCL12 has been found to play an essential role in promoting primordial germ cell transmigration through epithelial-like struc-

Chemokine Biology – Basic Research and Clinical Application, Volume I
edited by Bernhard Moser, Gordon L. Letts and Kuldeep Neote

tures, such as the hindgut epithelium in mouse and the endothelium in chick [5]. Of note, a possible role of interactions between CCR1 and its ligands in the initiation of trophoblastic invasion of maternal tissue has also been suggested [6]. The important role of chemokines in embryogenesis control represented the first evidence that chemokine receptors might also be expressed by resident cells in different tissues. Indeed, a large converging evidence has recognised the pivotal role of chemokines and their receptors in the biology of resident tissue cells largely beyond their chemotactic properties.

Chemokine receptors in epithelial tissues

Although chemokines were originally defined as host defense proteins and their main role is leukocyte recruitment, they and their receptors have other biological actions. Furthermore, many environmental stimuli of host of pathogen origin may lead to the induction of inflammatory chemokines expression and production in tissue cell types.

The expression of multiple chemokines in inflamed tissues, such as in the synovial lining cells of rheumatoid joints [7], autoimmune lesions in multiple sclerosis [8], ulcerative colitis and Crohn's disease [9], lung inflammation [10], sarcoidosis [11] and asthma [12], and the vascular inflammation that characterises arteriosclerosis [13], is well documented. Several receptors for inflammatory chemokines, CCR1, CCR2, CCR5 and CXCR3 in particular, are regularly detected in such lesions, while the expression of CCR3 tends to be restricted to allergic pathologies and the IL-8 receptors, CXCR1 and CXCR2, are more frequent in acute inflammation.

However, a great number of *in vivo* and *in vitro* studies demonstrated also the constitutive expression of chemokine receptors by resident epithelial cells of different tissues. The pattern of chemokines and chemokine receptors expression in epithelial tissues is summarised in Table 1.

Chemokines affecting vasculature-associated pericytes

Several studies have shown that the pericytes, smooth muscle-like mural cells that coat the wall of microvessels and are responsible for tissue fibrosis, may both express chemokines and be targets of the chemokine action [27–30]. In fact, pericytes express chemokine receptors, which, upon activation, elicit biologic actions that favour the processes of wound healing, including proliferation, migration, and extracellular matrix synthesis [31–33].

Human vascular smooth muscle cells (SMCs) express CCR2 [34], which makes these cells a likely target for CCL2. In fact, CCL2 can enhance the expression of

Table 1 - Expression and function of chemokine receptors in epithelial tissue cells

Type of tissue cells	Chemokine receptors	Function
Keratinocytes	CCR3 [14]	Inflammatory modulation
	CXCR1/CXCR2 [15]	Chemotaxis and proliferation
	CXCR3 [16]	Chemotaxis
Bronchial epithelial cells	CCR2 [17]	Proliferation and healing
	CCR3 [18]	Epithelial cell migration and proliferation
	CXCR4 [19]	Inflammatory modulation
Intestinal epithelium	CCR5 [20]	Cell migration
	CCR6 [21]	Cell migration, maintenance and renewal of the epithelium
	CXCR4 [20]	Hepatocytes
Ductular epithelial cell	CXCR4 [22]	Apoptosis
	CX3CR1 [23]	Wound healing response
Ectocervical epithelial cells	CCR5 [24]	Potential targets of HIV-1 infection
Podocyte	CCR4, CCR8, CCR9, [25]	Release of oxygen radicals
	CCR10, CXCR1-CXCR5 [25]	Release of oxygen radicals
	CXCR3 [26]	Induction of nephrin and podocin

integrins [35] as well as tissue factor [36] on SMCs. More recent findings [37] suggest that CCL2 can also directly induce SMC proliferation by stimulating the binding activity of activator protein 1. Cultured human arterial SMCs possess CCR5 at both mRNA and protein levels [33]. CCR5 on SMCs is functionally coupled, responding to CCL4 with increases in intracellular calcium concentration and tissue factor activity. CCR5 and CCL4 were also detected in SMCs of the atherosclerotic arterial wall, where they may play a role in mediating the inflammatory and prothrombotic responses associated with atherosclerosis. On the contrary, as determined by RT-PCR, human aortic SMCs do not express mRNA for other CCRs, including CCR1 [38], CCR3 [39], CCR4 [40], and DARC [41]. CXCL10 has been shown to act as a mitogen and chemoattractant for SMCs. Moreover, SMCs express CXCL10 in response to IL-1β and TNF-α in conjunction with IFN-γ and also in response to vascular injury, suggesting a role in pathogenesis of vascular diseases and injury [28].

Hepatic stellate cells (HSCs) and glomerular mesangial cells (MCs) are tissue-specific pericytes involved in tissue repair, a process that is regulated by chemokines.

In MCs expression of CCL2, CCL5, CXCL8, and CXCL10 has been repeatedly demonstrated [41–49]. CCL2 is rapidly upregulated in mouse, rat and human MCs after their activation by a variety of stimuli [41–43]. CCL5 is expressed 2 h after TNF-α stimulation by mouse MC [44] and it is also found to be expressed by primary human MCs [45]. CXCL8 is expressed by rat and human MCs [46, 47] and the expression of CXCL10 mRNA has been described for both mouse and human MCs [48, 49].

The expression of the chemokine receptor CXCR3 on human MCs was first reported by Romagnani P. and colleagues [31]. High expression of this receptor by MCs was seen by immunohistochemistry in kidney biopsies from patients with glomerulonephritis, characterised by resident mesangial cell proliferation, such as IgA nephropathy, membranoproliferative glomerulonephritis or rapidly progressive glomerulonephritis (also defined as "proliferative glomerulonephrites"). Moreover, CXCR3 was also found on the surface of cultured human MC (HMC), and appeared to mediate both intracellular Ca^{2+} influx and cell proliferation [50]. Furthermore, it was found that in both HMC and other types of vascular pericytes, CXCL10 and CXCL9 also induce chemotaxis and CXCR3 triggering results in Src activation, which in turn leads to the recruitment of Ras and activation of the ERK cascade [50]. In parallel, activation of PI 3-K and Akt can also be observed [50]. Taken all together, these findings may account for at least some mechanisms involved in the pathogenesis of proliferative GN.

Constitutive expression of the chemokine CCL21 on human podocytes and of its corresponding receptor CCR7 on MCs was also shown by immunohistochemistry of human kidney and these findings were confirmed in cultured cells and isolated glomeruli [51]. CCL21 has a positive effect on the proliferation and migration of MCs and leads to increased cell survival in Fas-induced apoptosis of human MC [51]. Moreover, activation of CCR7 on MCs by CCL21 enhances the degree and firmness of cell adhesion and increases cell spreading and the formation of cell–cell contacts, including integrin-linked kinase activation and F-actin rearrangements [52].

Inducible expression of the chemokine receptor CCR1 by human MCs after stimulation with a combination of the proinflammatory cytokines TNF-α, IL-1β and IFN-γ, has also been described [32]. In contrast to the effects observed with the ligands for CCR7 and CXCR3, stimulation of MCs with the CCR1 ligand CCL5 had no effect on cell proliferation and apoptosis. In conclusion, local chemokine generation and chemokine receptor expression on MCs may play an important role in the maintenance of glomerular homeostasis and in local remodelling processes.

HSCs express and secrete several CC chemokines, including CCL2 and CCL3 [53, 54]. Several lines of evidence indicate that CCL2 plays a role in the recruitment and maintenance of the inflammatory infiltrate during liver injury. CCL2 secretion is upregulated during chronic hepatitis and correlates with the number of cells infiltrating the portal tract [55]. *In vitro* and *in vivo* data indicate that HSCs may contribute to the expression of CCL2 within the liver during both chronic and acute injury [53, 54, 56]. On cultured human HSCs, CCL2 stimulates migration in a dose-dependent fashion and activates intracellular signalling, such as increase in cytosolic calcium concentration, PI3-K activity, protein tyrosine phosphorylation [56]. Cultured HSCs express functional CCR7, the activation of which stimulates cell migration and accelerates wound healing in an *in vitro* model. Exposure of HSCs to CCL21 triggered several signalling pathways, including extracellular signal-regulated kinase, Akt, and nuclear factor κB, resulting in induction of proinflammatory genes [57]. HSCs express CCR5, as shown by flow-cytometric analysis and RT-PCR [57], and respond to CCL5 with an increase in both intracellular calcium concentration and free radical formation. Furthermore, CCL5 induced ERK phosphorylation and HSC proliferation. Additionally, CCL5 induced focal adhesion kinase phosphorylation and a substantial increase in HSC migration [58]. HSC expressed functional CXCR3 receptors on the cells surface, and interaction with CXCR3 ligands resulted in increased chemotaxis, but not proliferation, through the Ras/ERK signalling cascade. Activation of CXCR3 stimulated Src phosphorylation and kinase activity and increased the activity of PI3-K [50].

Chemokines control of angiogenesis and wound healing

Tissue repair

Models of skin wound healing mimic inflammatory reactions that might also be relevant to infectious processes in general [59]. In this model, the interplay of CXC chemokines with growth factors, cytokines and adhesion molecules not only influences the sequential participation of inflammatory cells but, more importantly, regulates the inflammatory reaction leading to angiogenesis, tissue repair and new tissue generation [59, 60]. The repair process is initiated immediately after injury of blood vessels through the release from degranulating platelets of various growth factors, such as vascular endothelial growth factor (VEGF)-A, platelet-derived growth factor (PDGF), and several chemokines in large quantities. CXCL1, CXCL5 and CXCL7 initiate the neutrophil recruitment [59, 61, 62], whereas high amounts of CXCL4 contribute to the formation of blood clots [63]. This provides a barrier against invading microorganisms and serves as a matrix for the attachment of inflammatory cells that are recruited to wound tissue within a few hours of injury. The initial vessel-associated expression of CXCL1 facilitates neutrophil diapedesis [64]. Subsequently, the cooperative expression of CXCL1 and CXCL8 in the superficial wound bed supports additional neutrophil migration to the wound surface [65]. Neutrophils produce a wide variety of proteinases and reactive oxygen species as a defense against contaminating microorganisms and they are involved in the phagocytosis of cell debris. CXCR2 is also expressed on neovascularising ECs [65].

The time course of CXCL8 expression correlates with massive angiogenesis between days 1–4 [64], leading to the formation of new blood vessels. The newly formed connective tissue is known as granulation tissue because of the granular appearance of several capillaries. Accordingly, *CXCR2*-deficient mice exhibit a defective neutrophil recruitment, delayed monocyte recruitment and severe impairment of angiogenesis at the site of wounding [66]. Neutrophil accumulation is followed by the immigration of monocytes and macrophages, as a result of CCL2/CCR2 chemokine system [64, 67]. Interestingly, from days 0–6 after wounding, CXCL12 production by keratinocytes and fibroblasts is progressively downregulated, because of the inhibitory effect exerted by IL-1 and TNF. Given the ubiquitous expression of CXCR4 on both resident and inflammatory cell types, this probably represents a counter regulatory mechanism to avoid chronic inflammation [68]. High numbers of lymphocytes are also recruited during the whole period of healing and they represent the major leukocyte subpopulation on day 14. Between days 1–4, CXCL11, which is constitutively produced on the surface of human microvascular endothelial cells (HMVECs) [60] and is highly induced by epithelial monolayer disruption [64], contributes to the pronounced lymphocyte accumulation. Subsequently, CXCL9 and CXCL10, which are both T cell attractants [69, 70], are highly expressed at sites of lymphocyte accumulation [64]. Indeed, activated lymphocytes express high levels of CXCR3 [71]. The fact that vascularity increases until day 4, but remains constant afterwards, despite the presence of growth factors, such as bFGF and PDGF, suggests that the angiostatic properties of CXCL9 and CXCL10 can prevent unlimited vessel growth. In this context, the cell cycle dependence of CXCR3-B expression by HMVECs is of crucial importance [71]. Indeed, only ‘angiogenic’ ECs can respond to angiostatic stimuli, and therefore they arrest both migration and growth through inhibition mediated by CXCL11 present on the surface of adjacent ECs. This mechanism enables the generation of a finely regulated network of vessels (see below) without altering the properties and functions of quiescent ECs, which cannot respond to angiostatic chemokines. Finally, CXCL10, CXCL9 and CXCL11 mediate the migration of CXCR3-A-expressing pericytes and their proliferation around nascent vessels. The opposite effects of CXCL9, CXCL10 and CXCL11 on ECs and pericytes could be explained by distinct and sequential steps leading to angiogenesis. Of note, recruitment of pericytes occurs after the progression phase of angiogenesis that is determined by EC positioning and proliferation. The association of pericytes to newly formed blood vessels has been suggested to regulate endothelial cell proliferation, survival, migration, differentiation, and vascular branching. Therefore, these chemokines could contribute to vessel stabilisation by inhibiting cell cycle progression in ECs.

Migration and proliferation of keratinocytes at the wound edge are followed by the recruitment and proliferation of dermal fibroblasts. These cells subsequently acquire a contractile phenotype and transform into myofibroblasts, which have a major role in wound contraction. CXCL8 might directly stimulate re-epithelialisation, as a result of stimulating keratinocyte proliferation [72]. However, wound contraction is diminished by topical application of CXCL8, suggesting that elevated levels of this chemokine might also contribute to retarded wound repair [73]. Finally, a transition from granulation tissue to mature scar occurs, which is characterised by continued collagen synthesis and catabolism. CXCL10 and CXCL11 also deliver signals to the dermal compartment to synchronise the re-epithelialisation process. Indeed, these chemokines limit EGF-induced fibroblast motility, but promote the chemotaxis of undifferentiated keratinocytes [74]. A differentiated and strictly regulated CXCR3-A and CXCR3-B expression on keratinocytes and fibroblasts can be reasonably hypothesised and contributes to this pathway, but still needs to be proved. The possible roles of chemokines in the different steps of inflammatory processes from the starting tissue injury until the healing phase are summarised in Figure 1.

De novo blood vessel formation

Previous and more recent evidences indicate that ECs express specific receptors, which can account for an important role of chemokines in angiogenesis (Fig. 2A). Receptors for angiogenic chemokines expressed by ECs include CXCR1, CXCR2 and CXCR4 [75]. The first angiogenic chemokine receptor identified so far is CXCR4. *CXCR4/CXCL12*-deficient mice die prenatally and exhibit defects in the formation of gastrointestinal tract arteries, as well as defects in vessel development, haematopoiesis and cardiogenesis [1, 2]. The existence of a regulatory loop between VEGF-A and CXCL12/CXCR4 further supports the important role of this chemokine system in the regulation of angiogenesis. Indeed, CXCL12 upregulates VEGF-A production, and VEGF-A upregulates CXCR4 expression, thus generating an amplification circuit, which is critically influenced by hypoxia [76, 77]. Subsequently, the observation of angiogenesis impairment in *CXCR2*-deficient mice has allowed to demonstrate that this receptor mediates the angiogenic activity of CXCL1, CXCL2, CXCL3, CXCL5, CXCL6 and CXCL7.

The understanding of mechanisms responsible for CXC chemokine-mediated angiostatic effects (Fig. 2A) has been more difficult, mainly because CXCL4 and CXCL10 inhibit angiogenesis through both receptor-independent (i.e., competing with heparan sulfate proteoglycans on the cell surface or directly binding to these growth factors) and receptor-dependent mechanisms [78–80]. Recently, however, CXCR3 has been clearly detected in ECs, particularly at level of ECs from small vessels [81]. More importantly, it was found that CXCR3 expression by primary HMVECs was restricted to the S-phase of the cell cycle [81]. Our studies also led to the demonstration that CXCL11, the third known CXCR3-binding chemokine, was able to inhibit EC proliferation [81]. Furthermore, neutralising anti-CXCR3 antibodies blocked the antiproliferative activity induced on ECs by all three known

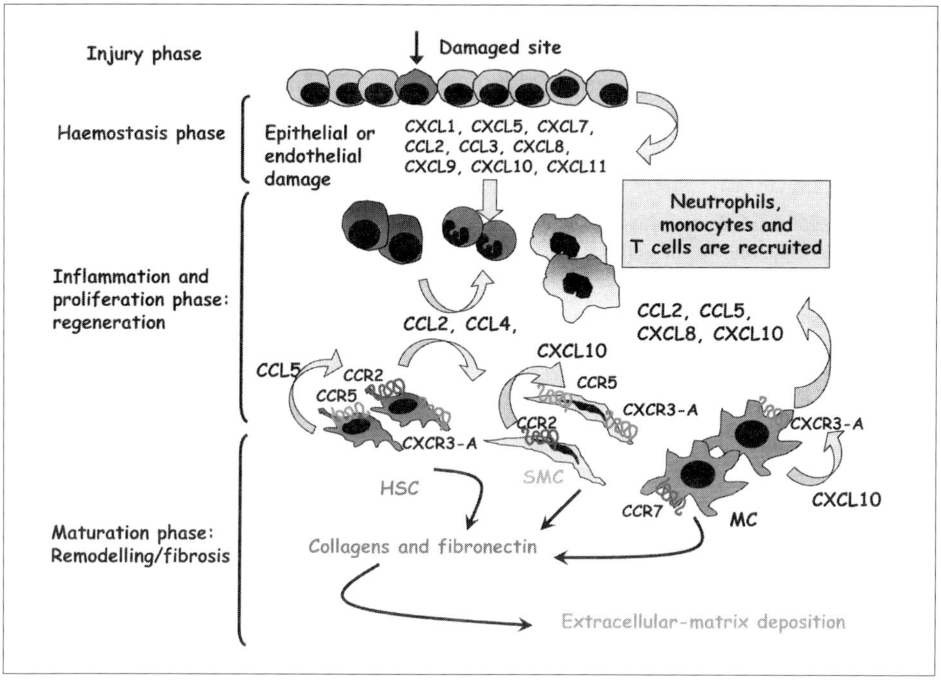

Figure 1
Role of chemokines in the different phases of inflammatory processes
In different tissues, the wound healing response shares many similarities, involving the recruitment of inflammatory cells and the deposition of extracellular matrix, to fill the gap created by the dying cells. Indeed, after tissue damage, chemokines such as CXCL1, CXCL5, CXCL7, CXCL8, CXCL9, CXCL10, CXCL11, CCL2, CCL3, lead to the recruitment of monocytes/macrophages, T cells and neutrophils. The concurrent presence of inflammation and extracellular matrix deposition is a characteristic of chronic tissue injury, where the persistence of a wound healing response may lead to permanent scarring and end-stage organ failure, such as in the case of glomerulosclerosis in the kidney, cirrhosis of the liver, atherosclerosis, or pulmonary fibrosis. The pivotal role played by vascular pericytes of different tissues in the process of wound healing has been clearly recognised in recent years. These cells become activated in the presence of damage to the specific tissue, proliferate, migrate, and acquire a myofibroblast-like phenotype, resulting in the production of extracellular matrix as part of the healing process. Pericytes responsible for tissue fibrosis may express chemokines such as CCL2, CCL4, CCL5, CXCL8, CXCL10, thus contributing to the pathogenesis of the inflammatory reaction. Furthermore, pericytes can also be targets of the action of chemokines, since they express chemokine receptors, such as CXCR3-A, CCR2, CCR5, CCR7, which, upon activation, elicit biologic actions that favour the wound healing process, including proliferation, migration, and extracellular matrix synthesis.

CXCR3 ligands, thus definitively proving that CXCR3 is the receptor involved in CXC chemokine-mediated angiostatic activity [81]. The role of CXCR3 in mediating the angiostatic activity of CXCL10 has recently been confirmed *in vivo* by blocking the angiostatic effects of CXCL10 in the rat cornea micropocket assay with a neutralising anti-CXCR3 antibody [82].

Some questions, however, still needed to be solved. First, the receptor for CXCL4, the most powerful angiostatic chemokine, remained unknown, despite the fact that this chemokine shares many activities with CXCL10. On the other side, CXCR3-binding chemokines also exhibit powerful chemotactic activity, whereas the CXCL4-mediated chemotactic effect is modest or absent [83]. Finally, the opposite effects exerted by CXCR3 ligands on HMVECs (inhibition of proliferation) and on vascular pericytes (increase of proliferation) [31, 84–86] allow to hypothesise the existence of cell-specific signal transduction pathways or even of distinct CXCR3 receptor variants.

Indeed, a distinct, previously unrecognised receptor, deriving from an alternative splicing of the *CXCR3* gene, was identified, which not only mediates the angiostatic activity of the three already known CXCR3 ligands, but also acts as functional receptor for CXCL4 [71]. By contrast, the known CXCR3, renamed CXCR3-A, mediated the proliferation of vascular pericytes in response to CXCL9, CXCL10 and CXCL11, whereas it bound CXCL4 with very low affinity [71]. Finally, monoclonal antibodies, that were selectively developed against CXCR3-B, reacted with ECs of different human tumour tissues but poorly, or not, with those from their normal counterparts, consistently with the previously described selective effects of both CXCL4 and CXCL10 on actively proliferating ECs [71]. Of note, another form of CXCL4 (CXCL4L1) has recently been isolated from thrombin-stimulated human platelets, which differed from CXCL4 in only three amino acids, and appeared to be more potent in inhibiting chemotaxis of HMVECs toward CXCL8 or bFGF [87]. Notably, a third variant of human CXCR3 (CXCR3-alt) resulting from alternative splicing via post-transcriptional exon skipping has also been identified [88]. However, the functional activity of this variant is not yet known.

Tumour formation

The course in angiogenesis usually correlates with the degree of infiltration by inflammatory leukocytes [59]. The coordination of angiogenesis and inflammation is due to the ability shared by ECs and leukocytes to respond to chemokines [61].

In physiologic processes, such as wound healing, the interplay of CXC chemokines with growth factors, cytokines and adhesion molecules regulates the events leading to angiogenesis. The repair process is initiated immediately after injury of blood vessels through the release of platelets-derived factors as described above. CXCL8 expression by wounded epithelial cells induces massive angiogene-

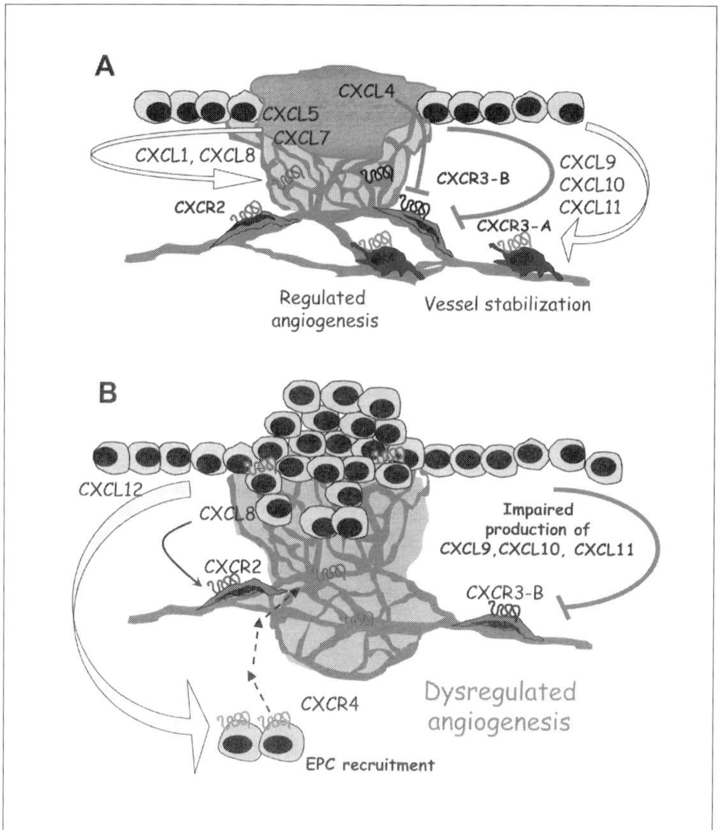

Figure 2
Role of chemokines in physiologic and dysregulated angiogenesis
(A) On wounding or tissue assault, platelets are activated and form a haemostatic plug, in which they release vasoactive mediators that regulate formation of the fibrin clot. CXCL1, CXCL5, CXCL7, derived from activated platelets, initiate the recruitment of neutrophils. Subsequently, CXCL8 expression by wounded epithelial cells induces massive angiogenesis, leading to the formation of new blood vessels that exhibit high CXCR2 expression. Conversely, expression of the angiostatic chemokines CXCL9, CXCL10, and CXCL11 prevents unlimited vessel growth, arresting migration and growth of proliferating endothelial cells, which selectively express CXCR3-B.
(B) An altered balance of CXC chemokines might be crucial in contributing to cancer development during chronic inflammatory processes through different mechanisms. Excessive production of angiogenic chemokines, such as CXCL8, and their receptor CXCR2, can lead to a level of inflammation that potentiates angiogenesis. Poor expression of angiostatic chemokines and of their receptor, CXCR3-B, can lead to a level of inflammation that potentiates angiogenesis or can directly alter the proliferative properties of resident epithelial cells.

sis, leading to the formation of new blood vessels expressing functional CXCR2 [64, 66]. Conversely, expression of the angiostatic chemokines CXCL9 and CXCL10 prevents unlimited vessel growth arresting migration and growth of proliferating ECs expressing CXCR3-B. CXCL10, CXCL9 and CXCL11 also mediate the migration of CXCR3-A-expressing pericytes and their proliferation around nascent vessels, thus determining their stabilisation.

On the other hand, tumours are described as "wounds that never heal" and appear to lack the appropriate balances between positive and negative control signals [89]. One of the main features of tumour blood vessels is their failure to become quiescent, enabling the constant growth of new tumour blood vessels [89]. Consequently, the tumour vasculature develops unique characteristics and becomes quite distinct from existing capillaries. Furthermore, the inappropriate or decreased vessel association with pericytes in tumours might account for both abnormal vessel diameters and sensitivity to VEGF inhibition [89].

Overexpression of angiogenic CXC chemokines favours the "tumour angiogenesis switch" and ultimately leads to tumour progression [89]. Lung colonisation and spontaneous metastasis in nude mice are inhibited by treatment with neutralising antibody against IL-8 [90]. Furthermore, CXCL8 expression in astrocytoma increases during tumour progression, due to reduced microenvironmental oxygen pressure and promotes angiogenesis by binding to CXCR2 [91]. CXCL8 and GRO-α are also induced by Kaposi Sarcoma Herpes Virus (KSHV) infection of endothelial cells and are crucial to the angiogenic phenotype developed by KSHV-infected ECs in cell culture and upon implantation into SCID mice [92]. A few data are available on the role of CXCL12 in angiogenesis progression in tumours. However, CXCL12 can contribute to tumour neovascularisation through vasculogenesis-mediated by EC precursors. Indeed, locally derived CXCL12 augments vasculogenesis and contributes to ischemic neovascularisation *in vivo* by augmenting the recruitment and survival of EC precursors [93]. Conversely, angiostatic chemokines play an important role in fighting tumour development and diffusion. Indeed, overexpression of CXCL4 and CXCL10 blocks tumour progression and can also induce regression of metastasis [94, 95]. The possibility that inadequate expression of CXCR3-B by angiogenic ECs during a chronic inflammatory process might favour the "tumour angiogenesis switch" might also be hypothesised. In 40 patients affect-

Resident epithelial cells undergo neoplastic progression and then, following hypoxia, "turn on" the expression of CXCR4. The production of CXCL12 in sites, such as lymph nodes, bone marrow, liver, and lung, then facilitates their invasion and migration to secondary sites to form a productive metastatic lesion and also potentiates angiogenesis, through its interaction with CXCR4. On the other hand, impaired production of CXCL9, CXCL10 and CXCL11 and/or their receptor CXCR3-A can result in impaired recruitment and activation of inflammatory cells resulting in escape of the tumour from immune surveillance.

ed by non small cell lung cancer (NSCLC), we observed a significant inverse correlation between CXCR3-B mRNA expression and both tumour stage and rate of lymph node invasion (Lazzeri E et al. manuscript in preparation). An inverse correlation between CXCR3-B expression and angiogenesis was only observed among patients with localised tumours and without lymph node invasion, suggesting that the loss of angiogenesis regulation by CXCR3-B might favour NSCLC diffusion. Similar findings were found in patients with renal cell carcinoma (Lazzeri E et al., manuscript in preparation). Collectively, dysregulation of chemokine production and/or interaction of chemokines with their receptor(s) appear to play an important role in the growth of cancer and in the formation of metastases. Figure 2B shows the possible role of different chemokines in the dysregulation of angiogenesis which occurs in neoplastic processes.

Chemokines control of other tissue cells

Many cell types in the brain express chemokines and chemokine receptors even under homeostatic conditions, arguing for a role of these molecules in normal brain processes. It has indeed been shown that CXCL12 and CCR3-binding chemokines reversibly inhibit neuronal progenitor cell (NPC) proliferation in isolated cells, neurospheres, and in hippocampal slice cultures [96]. On the other hand, CX3CL1 has been found to be able to promote survival of NPCs [96].

Cells of the central nervous system

There is also growing evidence for the role of chemokines in the regulation of central nervous system (CNS) diseases. Elevated levels of chemokines have been indeed observed in both experimental autoimmune encephalomyelitis (EAE) and multiple sclerosis (MS), suggesting that these molecules act as regulators of brain inflammation [97, 98]. However, chemokines not only function as key mediators which promote leukocyte infiltration of demyelinating lesions in both EAE and MS, but they also act on microglia and astrocytes by inducing their migration to sites of inflammation, and their proliferation that could represent the basis of pathological conditions such as gliosis. The major receptors on these cells appear to be CXCR1 and CXCR3, but also CCR3 [99].

Osteoclasts

Although much has been learned of the mechanisms by which the migration and differentiation of osteoclasts (OCs) are induced, only recently the essential role of chemokines in this process has been recognised. CXCL12 stimulates matrix metalloproteinase-9 activity on pre-OCs, thus favouring their recruitment to sites for OC differentiation and bone readsorption [100]. On the other side, CXCL8 has been shown to play a direct effect on OC differentiation and activity by interacting with its specific receptor CXCR1, which appears to be expressed on the surface of these cells [101]. CCL9 and its receptor CCR1 have also been found on OCs, suggesting that this chemokine and its receptor may also play a role in the regulation of bone readsorption [102]. Moreover, high levels of CCL3 have been found in bone marrow samples from patients with multiple myeloma, suggesting that it may be one of the major factors responsible for the increased OC stimulatory activity in patients with this disease [103]. However, a more recent study, based on the use of gene array, showed that of all the mediators screened, CCL15 was the most strongly upregulated in stimulated OC precursors [104]. More importantly, neutralisation of CCL15 resulted in strongly reduced OC formation and reduced resorptive activity, since CCL15 also promoted OC survival and prevented OC apoptosis. These results suggest that OCs can protect themselves from apoptosis through production of CCL15 as an autocrine survival factor [104].

Conclusions

Chemokines are secretory proteins produced by leukocytes and tissue cells either constitutively or after induction, and exert their effects locally in paracrine or autocrine fashion via their binding to heptahelical G-protein coupled receptors. The increase in the secretion of chemokines during inflammation results in the selective recruitment of leukocytes into inflamed tissues such as skin, brain, lung, kidneys and gastrointestinal tract. In these organs many types of cells secrete chemokines, suggesting that, if the appropriate stimulus is given, most cells can secrete chemokines.

Moreover, in organs such as kidney, lung and liver, chemokines may play an important role in the maintenance of tissue homeostasis, in local remodelling processes and may modulate the progression of fibrosis by acting on tissue specific pericytes. Most importantly, chemokines have been found to have a main role in the regulation of angiogenesis and tumour-related immunity, and in promoting organ-specific metastases.

Our knowledge on the roles of chemokines in the pathophysiology of disease are derived from studies utilising animal models of disease and mice with deleted chemokine receptor genes. The main problems in studying the role of chemokines in these models might be represented by the great redundancy shown by the chemokine system (i.e., different chemokines can bind a single chemokine receptor and a single chemokine can bind more than a receptor) and some differences between species in the expression of chemokines and chemokine receptors and in their binding properties. However, there is growing evidence that the neutralisation of chemokine activity may have a therapeutic value. Indeed, chemokine analogues with antagonist or partial agonist activity proved effective in animal models as inhibitors of inflammatory pathologies. In particular, given the role of chemokines in excessive fibrosis, novel strategies aimed at preventing fibrotic disease will likely need to address the early engagement of inflammatory cells by tissue epithelial and interstitial cells, and possibly modulate the ability of resident tissue cells to generate and/or recognise profibrotic signals supplied by chemokines. Finally, understanding the biology of factors that contribute to cancer tumourigenicity, avoidance of host immunity, metastases and angiogenesis may lead to novel strategies for therapeutic intervention of this devastating disease.

References

1 McGrath KE, Koniski AD, Maltby KM, McGann JK, Palis J (1999) Embryonic expression and function of the chemokine SDF-1 and its receptor, CXCR4. *Developm Biol* 213: 442–456

2 Nagasawa T, Hirota S, Tachibana K, Takakura N, Nishikawa S-I, Kitamura Y, Yoshida N, Kikutani H, Kishimoto T (1996) Defects of B-cell lymphopoiesis and bone-marrow myelopoiesis in mice lacking the CXC chemokine PBSF/SDF1. *Nature* 382: 635–638

3 Tachibana K, Hirota S, Iizasa H, Yoshida H, Kawabata K, Kataoka Y, Kitamura Y, Matsushima K, Yoshida N, Nishikawa S-I et al (1998). The chemokine receptor CXCR4 is essential for vascularization of the gastrointestinal tract. *Nature* 393: 591–594

4 Grone H-J, Cohen CD, Grone E, Schmidt C, Kretzler M, Schlondorff D, Nelson PJ (2002) Spatial and temporally restricted expression of chemokines and chemokine receptors in the developing human kidney. *J Am Soc Nephrol* 13: 957–967

5 Stebler J, Spieler D, Slanchev K, Molyneaux KA, Richter U, Cojocaru V, Tarabykin V, Wylie C, Kessel M, Raz E (2004). Primordial germ cell migration in the chick and mouse embryo: the role of the chemokine SDF-1/CXCL12. *Developm Biol* 272: 351–361

6 Sato Y, Higuchi T, Yoshioka S, Tatsumi K, Fujiwara H, Fujii S (2003) Trophoblasts acquire a chemokine receptor, CCR1, as they differentiate towards invasive phenotype. *Development* 130: 5519–5532

7 Haringman JJ, Kraan MC, Smeets TJ, Zwinderman KH, Tak PP (2003) Chemokine blockade and chronic inflammatory disease: proof of concept in patients with rheumatoid arthritis. *Ann Rheum Dis* 62: 715–721

8 Dogan RN, Karpus WJ (2004) Chemokines and chemokine receptors in autoimmune encephalomyelitis as a model for central nervous system inflammatory disease regulation. *Front Biosci* 9: 1500–1505

9 MacDermott RP, Sanderson IR, Reinecker HC (1998) The central role of chemokines (chemotactic cytokines) in the immunopathogenesis of ulcerative colitis and Crohn's disease. *Inflamm Bowel Dis* 4: 54–67

10 D'Ambrosio D, Mariani M, Panina-Bordignon P, Sinigaglia F (2001) Chemokines and

their receptors guiding T lymphocyte recruitment in lung inflammation. *Am J Respir Crit Care Med* 164: 1266–1275

11 Agostini C, Meneghin A, Semenzato G (2002) T-lymphocytes and cytokines in sarcoidosis. *Curr Opin Pulm Med* 8: 435–440

12 Bisset LR, Schmid-Grendelmeier P (2005) Chemokines and their receptors in the pathogenesis of allergic asthma: progress and perspective. *Curr Opin Pulm Med* 11: 35–42

13 Boisvert WA (2004) Modulation of atherogenesis by chemokines *Trends Cardiovasc Med* 14: 161–165

14 Wakugawa M, Nakamura K, Akatsuka M, Kim SS, Yamada Y, Kawasaki H, Tamaki K, Furue M (2001) Expression of CC chemokine receptor 3 on human keratinocytes *in vivo* and *in vitro* –upregulation by RANTES. *J Dermatol Sci* 25: 229–235

15 Kulke R, Bornscheuer E, Schluter C, Bartels J, Rowert J, Sticherling M, Christophers E (1998) The CXC receptor 2 is overexpressed in psoriatic epidermis. *J Invest Dermatol* 110: 90–94

16 Satish L, Blair HC, Glading A, Wells A (2005) Interferon-inducible protein 9 (CXCL11)-induced cell motility in keratinocytes requires calcium flux-dependent activation of mu-calpain. *Mol Cell Biol* 25: 1922–1941

17 Lundien MC, Mohammed KA, Nasreen N, Tepper RS, Hardwick JA, Sanders KL, Van Horn RD, Antony VB (2002) Induction of MCP-1 expression in airway epithelial cells: role of CCR2 receptor in airway epithelial injury. *J Clin Immunol* 22: 144–152

18 Stellato C, Brummet ME, Plitt JR, Shahabuddin S, Baroody FM, Liu MC, Ponath PD, Beck LA (2001) Expression of the C–C chemokine receptor CCR3 in human airway epithelial cells. *J Immunol* 166: 1457–1461

19 Eddleston J, Christiansen SC, Zuraw BL (2002) Functional expression of the C-X-C chemokine receptor CXCR4 by human bronchial epithelial cells: regulation by proinflammatory mediators. *J Immunol* 169: 6445–6451

20 Dwinell MB, Eckmann L, Leopard JD, Varki NM, Kagnoff MF (1999) Chemokine receptor expression by human intestinal epithelial cells. *Gastroenterology* 117: 359–367

21 Yang CC, Ogawa H, Dwinell MB, McCole DF, Eckmann L, Kagnoff MF (2005) Chemokine receptor CCR6 transduces signals that activate p130Cas and alter cAMP-stimulated ion transport in human intestinal epithelial cells. *Am J Physiol Cell Physiol* 288: C321–C328

22 Vlahakis SR, Villasis-Keever A, Gomez TS, Bren GD, Paya CV (2003) Human immunodeficiency virus-induced apoptosis of human hepatocytes via CXCR4. *J Infect Dis* 188: 1455–1460

23 Efsen E, Grappone C, DeFranco RM, Milani S, Romanelli RG, Bonacchi A, Caligiuri A, Failli P, Annunziato F, Pagliai G et al (2002) Up-regulated expression of fractalkine and its receptor CX3CR1 during liver injury in humans. *J Hepatol* 37: 39–47

24 Patterson BK, Landay A, Andersson J, Brown C, Behbahani H, Jiyamapa D, Burki Z, Stanislawski D, Czerniewski MA, Garcia P (1998) Repertoire of chemokine receptor expression in the female genital tract: implications for human immunodeficiency virus transmission. *Am J Pathol* 153: 481–490

25 Huber TB, Reinhardt HC, Exner M, Burger JA, Kerjaschki D, Saleem MA, Pavenstadt H (2002) Expression of functional CCR and CXCR chemokine receptors in podocytes. *J Immunol* 168: 6244–6252

26 Han GD, Koike H, Nakatsue T, Suzuki K, Yoneyama H, Narumi S, Kobayashi N, Mundel P, Shimizu F, Kawachi H (2003) IFN-inducible protein-10 has a differential role in podocyte during Thy 1.1 glomerulonephritis. *J Am Soc Nephrol* 14: 3111–3126

27 Gharaee-Kermani M, Denholm EM, Phan SH (1996) Costimulation of fibroblast collagen and transforming growth factor beta1 gene expression by monocyte chemoattractant protein-1 via specific receptors. *J Biol Chem* 271: 17779–17784

28 Wang X, Yue TL, Ohlstein EH, Sung CP, Feuerstein GZ (1996) Interferon-inducible protein-10 involves vascular smooth muscle cell migration, proliferation, and inflammatory response. *J Biol Chem* 271: 24286–24293

29 Schecter AD, Rollins BJ, Zhang YJ, Charo IF, Fallon JT, Rossikhina M, Giesen PL, Nemerson Y, Taubman MB (1997) Tissue factor is induced by monocyte chemoattractant protein-1 in human aortic smooth muscle and THP-1 cells. *J Biol Chem* 272: 28568–28573

30 Marra F, Romanelli RG, Giannini C, Failli P, Pastacaldi S, Arrighi MC, Pinzani M, Laffi G, Montalto P, Gentilini P (1999) Monocyte chemotactic protein-1 as a chemoattractant for human hepatic stellate cells. *Hepatology* 29: 140–148

31 Romagnani P, Beltrame C, Annunziato F, Lasagni L, Luconi M, Galli G, Cosmi L, Maggi E, Salvadori M, Pupilli C et al (1999) Role for interactions between IP-10/Mig and CXCR3 in proliferative glomerulonephritis. *J Am Soc Nephrol* 10: 2518–2525

32 Banas B, Luckow B, Moller M, Klier C, Nelson PJ, Schadde E, Brigl M, Halevy D, Holthofer H, Reinhart B et al (1999) Chemokine and chemokine receptor expression in a novel human mesangial cell line. *J Am Soc Nephrol* 10: 2314–2322

33 Schecter AD, Calderon TM, Berman AB, McManus CM, Fallon JT, Rossikhina M, Zhao W, Christ G, Berman JW, Taubman MB (2000) Human vascular smooth muscle cells possess functional CCR5. *J Biol Chem* 275: 5466–5471

34 Hayes IM, Jordan NJ, Towers S, Smith G, Paterson JR, Earnshaw JJ, Roach AG, Westwick J, Williams RJ (1998) Human vascular smooth muscle cells express receptors for CC chemokines. *Arterioscler Thromb Vasc Biol* 18: 397–403

35 Ikeda U, Ikeda M, Seino Y, Takahashi M, Kasahara T, Kano S, Shimada K (1993) Expression of intercellular adhesion molecule-1 on rat vascular smooth muscle cells by pro-inflammatory cytokines. *Atherosclerosis* 104: 61–68

36 Schecter AD, Rollins BJ, Zhang YJ, Charo IF, Fallon JT, Rossikhina M, Giesen PL, Nemerson Y, Taubman MB (1997) Tissue factor is induced by monocyte chemoattractant protein-1 in human aortic smooth muscle and THP-1 cells. *J Biol Chem* 272: 28568–28573

37 Wang N, Tabas I, Winchester R, Ravalli S, Rabbani LE, Tall A (1996) Interleukin 8 is induced by cholesterol loading of macrophages and expressed by macrophage foam cells in human atheroma. *J Biol Chem* 271: 8837–8842

38 Gao JL, Kuhns DB, Tiffany HL, McDermott D, Li X, Francke U, Murphy PM (1993)

Structure and functional expression of the human macrophage inflammatory protein 1 alpha/RANTES receptor. *J Exp Med* 177: 1421–1427
39 Kitaura M, Nakajima T, Imai T, Harada S, Combadiere C, Tiffany HL, Murphy PM, Yoshie O (1996) Molecular cloning of human eotaxin, an eosinophil-selective CC chemokine, and identification of a specific eosinophil eotaxin receptor, CC chemokine receptor 3. *J Biol Chem* 271: 7725–7730
40 Youn BS, Kim SH, Lyu MS, Kozak CA, Taub DD, Kwon BS (1997) Molecular cloning and characterization of a cDNA, CHEMR1, encoding a chemokine receptor with a homology to the human C-C chemokine receptor, CCR-4. *Blood* 89: 4448–4460
41 Hora K, Satriano JA, Santiago A, Mori T, Stanley ER, Shan Z, Schlöndorff D (1992) Receptors for IgG complexes activate synthesis of monocyte chemoattractant peptide 1 and colony-stimulating factor 1. *Proc Natl Acad Sci USA* 89: 1745–1749
42 Pai R, Ha H, Kirschenbaum MA, Kamanna VS (1996) Role of tumor necrosis factor-on mesangial cell MCP-1 expression and monocyte migration: Mechanisms mediated by signal transduction. *J Am Soc Nephrol* 7: 914–923
43 Largen PJ, Tam FW, Rees AJ, Cattell V (1995). Rat mesangial cells have a selective role in macrophage recruitment and activation. *Exp Nephrol* 3: 34–39
44 Wolf G, Aberle S, Thaiss F, Nelson PJ, Krensky AM, Neilson EG, Stahl RA (1993) TNF alpha induces expression of the chemoattractant cytokine RANTES in cultured mouse mesangial cells. *Kidney Int* 44: 795–804
45 Schwarz M, Radeke HH, Resch K, Uciechowski P (1997) Lymphocyte-derived cytokines induce sequential expression of monocyte- and T cell-specific chemokines in human mesangial cells. *Kidney Int* 52: 1521–1531
46 Brown Z, Strieter RM, Chensue SW, Ceska M, Lindley I, Neild GH, Kunkel SL, Westwick J (1991) Cytokine-activated human mesangial cells generate the neutrophil chemoattractant, interleukin 8. *Kidney Int* 40: 86–90
47 Robson RL, Westwick J, Brown Z (1995) Interleukin-1-induced IL-8 and IL-6 gene expression and production in human mesangial cells is differentially regulated by cAMP. *Kidney Int* 48: 1767–1777
48 Gomez Chiarri M, Hamilton TA, Egido J, Emancipator SN (1993) Expression of IP-10, a lipopolysaccharide- and interferon-gamma-inducible protein, in murine mesangial cells in culture. *Am J Pathol* 142: 433–439
49 Duque N, Gomez Guerrero C, Egido J (1999) Interaction of IgA with Fc alpha receptors of human mesangial cells activates transcription factor nuclear factor- B and induces expression and synthesis of monocyte chemoattractant protein-1, IL-8, and IFN-inducible protein 10. *J Immunol* 159: 3474–3482
50 Bonacchi A, Romagnani P, Romanelli RG, Efsen E, Annunziato F, Lasagni L, Francalanci M, Serio M, Laffi G, Pinzani M et al (2001) Signal transduction by the chemokine receptor CXCR3. *J Biol Chem* 276: 9945–9954
51 Banas B, Wörnle M, Berger T, Nelson PJ, Cohen CD, Kretzler M, Pfirstinger J, Mack M, Lipp M, Gröne HJ et al (2002) Roles of SLC/CCL21 and CCR7 in human kidney

for mesangial proliferation, migration, apoptosis and tissue homeostasis. *J Immunol* 168: 4301–4307

52 Banas B, Wornle M, Merkle M, Gonzalez-Rubio M, Schmid H, Kretzler M, Pietrzyk MC, Fink M, de Lema GP, Schlondorff D (2004) Binding of the chemokine SLC/CCL21 to its receptor CCR7 increases adhesive properties of human mesangial cells. *Kidney Int* 66: 2256–2263

53 Marra F, Valente AJ, Pinzani M, Abboud HE (1993) Cultured human liver fat-storing cells produce monocyte chemotactic protein-1. Regulation by proinflammatory cytokines. *J Clin Invest* 92: 1674–1680

54 Czaja MJ, Geerts A, Xu J, Schmiedeberg P, Ju Y (1994) Monocyte chemoattractant protein 1 (MCP-1) expression occurs in toxic rat liver injury and human liver disease. J Leukoc Biol 55: 120–126

55 Marra F, DeFranco R, Grappone C, Milani S, Pastacaldi S, Pinzani M, Romanelli RG, Laffi G, Gentilizi P (1998) Increased expression of Monocyte Chemotactic Protein-1 during active hepatic fibrogenesis: Correlation with monocyte infiltration. *Am J Pathol* 152: 423–430

56 Marra F, Romanelli RG, Giannini C, Failli P, Pastacaldi S, Arrighi MC, Pinzani M, Laffi G, Montalto P, Gentilini P (1999) Monocyte chemotactic protein-1 as a chemoattractant for human hepatic stellate cells. *Hepatology* 29:140–148

57 Bonacchi A, Petrai I, Defranco RM, Lazzeri E, Annunziato F, Efsen E, Cosmi L, Romagnani P, Milani S, Failli P et al (2003) The chemokine CCL21 modulates lymphocyte recruitment and fibrosis in chronic hepatitis C. *Gastroenterology* 125: 1060–1076

58 Schwabe RF, Bataller R, Brenner DA (2003) Human hepatic stellate cells express CCR5 and RANTES to induce proliferation and migration. *Am J Physiol Gastrointest Liver Physiol* 285: G949–G958

59 Griffioen AW, Molema G (2000) Angiogenesis: potentials for pharmacologic intervention in the treatment of cancer, cardiovascular diseases, and chronic inflammation. *Pharmacol Rev* 52: 237–268

60 Spinetti G, Camarda G, Bernardini G, Romano Di Peppe S, Capogrossi MC, Napolitano M (2001) The chemokine CXCL13 (BCA-1) inhibits FGF-2 effects on endothelial cells. *Biochem Biophys Res Commun* 289: 19–24

61 Werner S, Grose R (2003) Regulation of wound healing by growth factors and cytokines. *Physiol Rev* 83: 835–870

62 Gillitzer R, Goebeler M (2001) Chemokines in cutaneous wound healing. *J Leukoc Biol* 69: 513–521

63 Shuman MA, Levine SP (1978) Thrombin generation and secretion of platelet Factor 4 during blood clotting. *J Clin Invest* 61: 1102–1106

64 Engelhardt E, Toksoy A, Goebeler M, Debus S, Brocker EB, Gillitzer R (1998) Chemokines IL-8, GROalpha, MCP-1, IP-10, and Mig are sequentially and differentially expressed during phase-specific infiltration of leukocyte subsets in human wound healing. *Am J Pathol* 153: 1849–1860

65 Kemeny L, Szolnoky G, Kenderessy AS, Gyulai R, Kiss M, Michel G, Nagy K, Ruzicka

T, Dobozy A (1994) Role of interleukin-8 receptor in skin. *Int Arch Allergy Immunol* 104: 317–322

66 Devalaraja RM, Nanney LB, Du J, Qian Q, Yu Y, Devalaraja MN, Richmond A (2000) Delayed wound healing in CXCR2 knockout mice. *J Invest Dermatol* 115: 234–244

67 Dipietro LA, Reintjes MG, Low QE, Levi B, Gamelli RL (2001) Modulation of macrophage recruitment into wounds by monocyte chemoattractant protein-1. *Wound Repair Regen* 9: 28–33

68 Fedyk ER, Jones D, Critchley HO, Phipps RP, Blieden TM, Springer TA (2001) Expression of stromal-derived factor-1 is decreased by IL-1 and TNF in dermal wound healing. *J Immunol* 166: 5749–5755

69 Zlotnik A, Yoshie O (2000) Chemokines: a new classification system and their role in immunity. *Immunity* 12: 121–127

70 Rossi D, Zlotnik A (2000) The biology of chemokines and their receptors. *Ann Rev Immunol* 18: 217–242

71 Lasagni L, Francalanci M, Annunziato F, Lazzeri E, Giannini S, Cosmi L, Sagrinati C, Mazzinghi B, Orlando C, Maggi E et al (2003) An alternatively spliced variant of CXCR3 mediates the IP-10, Mig and I-TAC induced-inhibition of endothelial cell growth and acts as functional receptor for PF-4. *J Exp Med* 197: 1537–1549

72 Iocono JA, Colleran KR, Remick DG, Gillespie BW, Ehrlich HP, Garner WL (2000) Interleukin-8 levels and activity in delayed-healing human thermal wounds. *Wound Repair Regen* 8: 216–225

73 Rennekampff HO, Hansbrough JF, Kiessig V, Dore C, Sticherling M, Schroder JM (2000) Bioactive interleukin-8 is expressed in wounds and enhances wound healing. *J Surg Res* 93: 41–54

74 Shiraha H, Glading A, Gupta K, Wells A (1999) IP-10 inhibits epidermal growth factor-induced motility by decreasing epidermal growth factor receptor-mediated calpain activity. *J Cell Biol* 146: 243–254

75 Salcedo R, Oppenheim JJ (2003) Role of chemokines in angiogenesis: CXCL12/SDF-1 and CXCR4 interaction, a key regulator of endothelial cell responses. *Microcirculation* 10: 359–370

76 Salcedo R, Wasserman K, Young HA, Grimm MC, Howard OM, Anver MR, Kleinman HK, Murphy WJ, Oppenheim JJ (1999) Vascular endothelial growth factor and basic fibroblast growth factor induce expression of CXCR4 on human endothelial cells: *in vivo* neovascularization induced by stromal-derived factor-1alpha. *Am J Pathol* 154: 1125–1135

77 Staller P, Sulitkova J, Lisztwan J, Moch H, Oakeley EJ, Krek W (2003) Chemokine receptor CXCR4 downregulated by Von Hippel-Lindau tumor suppressor pVHL. *Nature* 425: 307–311

78 Sulpice E, Bryckaert M, Lacour J, Contreres JO, Tobelem G (2002) Platelet factor 4 inhibits FGF2-induced endothelial cell proliferation via the extracellular signal-regulated kinase pathway but not by the phosphatidylinositol 3-kinase pathway. *Blood* 100: 3087–3094

79 Gentilini G, Kirschbaum NE, Augustine JA, Aster RH, Visentin GP (1999) Inhibition of human umbilical vein endothelial cell proliferation by the CXC chemokine, platelet factor 4 (PF-4), is associated with impaired downregulation of $p21^{Cip1/WAF1}$. *Blood* 93: 25–33
80 Jouan V, Canron X, Alemany M, Caen JP, Quentin G, Plouet J, Bikfalvi A (1999) Inhibition of *in vitro* angiogenesis by platelet factor-4-derived peptides and mechanism of action. *Blood* 94: 984–993
81 Romagnani P, Annunziato F, Lasagni L, Lazzeri E, Beltrame C, Francalanci M, Uguccioni M, Galli G, Cosmi L, Maurenzig L et al (2001) Cell cycle-dependent expression of CXC chemokine receptor 3 by endothelial cells mediates angiostatic activity. *J Clin Invest* 107: 53–63
82 Strieter RM, Belperio JA, Phillips RJ, Keane MP (2004) CXC chemokines in angiogenesis of cancer. *Semin Cancer Biol* 14: 195–200
83 Zucker MB, Katz IR (1991) Platelet factor 4: production, structure, and physiologic and immunologic action. *Proc Soc Exp Biol Med* 198: 693–702
84 Romagnani P, Lazzeri E, Lasagni L, Mavilia C, Beltrame C, Francalanci M, Rotondi M, Annunziato F, Maurenzig L, Cosmi L (2002) IP-10 and Mig production by glomerular cells in human proliferative glomerulonephritis and regulation by nitric oxide. *J Am Soc Nephrol* 13: 53–64
85 Zhao DX, Hu Y, Miller GG, Luster AD, Mitchell RN, Libby P (2002) Differential expression of the IFN-{gamma}-inducible CXCR3-binding chemokines, IFN-inducible protein 10, monokine induced by IFN, and IFN-inducible T Cell {alpha} chemoattractant in human cardiac allografts: association with cardiac allograft vasculopathy and acute rejection. *J Immunol* 169: 1556–1560
86 Wang X, Yue TL, Ohlstein EH, Sung CP, Feuerstein GZ (1996) Interferon-inducible protein-10 involves vascular smooth muscle cell migration, proliferation, and inflammatory response. *J Biol Chem* 271: 24286–24293
87 Struyf S, Burdick MD, Proost P, Van Damme J, Strieter RM (2004) Platelets release CXCL4L1, a nonallelic variant of the chemokine platelet factor 4/CXCL4 and potent inhibitor of angiogenesis. *Circulation Res* 95: 855–857
88 Ehlert JE, Addison CA, Burdick MD, Kunkel SL, Strieter RM (2004) Identification and partial characterization of a variant of human CXCR3 generated by posttranscriptional exon skipping. *J Immunol* 173: 6234–6240
89 Bergers G, Benjamin LE (2003) Tumorigenesis and the angiogenic switch. *Nat Rev Cancer* 3: 401–410
90 Rofstad EK, Halsor ER (2000) Vascular endothelial growth factor, interleukin 8, platelet-derived endothelial cell growth factor, and basic fibroblast growth factor promote angiogenesis and metastasis in human melanoma xenografts. *Cancer Res* 60: 4932–4938
91 Desbaillets I, Diserens AC, Tribolet N, Hamou NF, Van Meir EG (1997) Upregulation of interleukin 8 by oxygen-deprived cells in glioblastoma suggests a role in leukocyte activation, chemotaxis, and angiogenesis. *J Exp Med* 186: 1201–1212

92 Lane BR, Liu J, Bock PJ, Schols D, Coffey MJ, Strieter RM, Polverini PJ, Markovitz DM (2002) Interleukin-8 and growth-regulated oncogene α mediate angiogenesis in Kaposi's sarcoma. *J Virol* 76: 11570–11583
93 Yamaguchi J, Kusano KF, Masuo O, Kawamoto A, Silver M, Murasawa S, Bosch-Marce M, Masuda H, Losordo DW, Isner JM et al (2003) Stromal cell-derived factor-1 effects on *ex vivo* expanded endothelial progenitor cell recruitment for ischemic neovascularization. *Circulation* 107: 1322–1328
94 Tanaka T, Manome Y, Wen P, Kufe DW, Fine HA (1997) Viral vector-mediated transduction of a modified platelet factor 4 cDNA inhibits angiogenesis and tumor growth. *Nat Med* 3: 437–442
95 Homey B, Muller A, Zlotnik A (2002) Chemokines: agents for the immunotherapy of cancer? *Nat Rev Immunol* 2: 175–184
96 Krathwohl MD, Kaiser JL (2004) Chemokines promote quiescence and survival of human neural progenitors. *Stem Cells* 22: 109–118
97 Glabinski AR, Ransohoff RM (1999) Chemokines and chemokine receptors in CNS pathology. *J Neurovirol* 5: 3–12
98 Zhang L, He T, Talal A, Wang G, Framkel SS, Ho DD (2000) Chemokines and chemokine receptors in the pathogenesis of multiple sclerosis. *Mult Scler* 6: 3–13
99 Flynn G, Maru S, Loughlin J, Romero JA, Male D (2003) Regulation of chemokine receptor expression in human microglia and astrocytes. *J Neuroimmunol* 136: 84–93
100 Yu X, Collin-Osdoby P, Osdopy P (2003) SDF-1 increases recruitment of osteoclast precursors by upregulation of matrix metalloproteinase-9 activity. *Connect Tissue Res* 44 suppl 1: 79–84
101 Bendre MS, Montague DC, Peery T, Akel NS, Gaddy, D, Suva LJ (2003) Interleukin-8 stimulation of osteoclastogenesis and bone resorption is a mechanism for the increased osteolysis of metastatic bone disease. *Bone* 33: 28–37
102 Lean JM, Murphy C, Fuller K, Chambers TJ (2002) CCL9/MIP-1γ and its receptor CCR1 are the major chemokine ligand/receptor species expressed by osteoclasts. *J Cell Biochem* 87: 386–393
103 Choi SJ, Cruz JC, Craig F, Chung H, Devlin RD, Roodman GD, Alsina M (2000) Macrophage inflammatory protein 1α is a potential osteoclast stimulatory factor in multiple myeloma. *Blood* 15: 671–675
104 Okamatsu Y, Kim D, Battaglino R, Sasaki H, Spate U, Stashenko P (2004) MIP-1γ promotes receptor activator of NF-κB ligand-induced osteoclast formation and survival. *J Immunol* 173: 2084–2090

Index

The PIR-Series
Progress in Inflammation Research

Homepage: http://www.birkhauser.ch

Up-to-date information on the latest developments in the pathology, mechanisms and therapy of inflammatory disease are provided in this monograph series. Areas covered include vascular responses, skin inflammation, pain, neuroinflammation, arthritis cartilage and bone, airways inflammation and asthma, allergy, cytokines and inflammatory mediators, cell signalling, and recent advances in drug therapy. Each volume is edited by acknowledged experts providing succinct overviews on specific topics intended to inform and explain. The series is of interest to academic and industrial biomedical researchers, drug development personnel and rheumatologists, allergists, pathologists, dermatologists and other clinicians requiring regular scientific updates.

Available volumes:

T Cells in Arthritis, P. Miossec, W. van den Berg, G. Firestein (Editors), 1998
Chemokines and Skin, E. Kownatzki, J. Norgauer (Editors), 1998
Medicinal Fatty Acids, J. Kremer (Editor), 1998
Inducible Enzymes in the Inflammatory Response,
 D.A. Willoughby, A. Tomlinson (Editors), 1999
Cytokines in Severe Sepsis and Septic Shock, H. Redl, G. Schlag (Editors), 1999
Fatty Acids and Inflammatory Skin Diseases, J.-M. Schröder (Editor), 1999
Immunomodulatory Agents from Plants, H. Wagner (Editor), 1999
Cytokines and Pain, L. Watkins, S. Maier (Editors), 1999
In Vivo *Models of Inflammation*, D. Morgan, L. Marshall (Editors), 1999
Pain and Neurogenic Inflammation, S.D. Brain, P. Moore (Editors), 1999
Anti-Inflammatory Drugs in Asthma, A.P. Sampson, M.K. Church (Editors), 1999
Novel Inhibitors of Leukotrienes, G. Folco, B. Samuelsson, R.C. Murphy (Editors), 1999
Vascular Adhesion Molecules and Inflammation, J.D. Pearson (Editor), 1999
Metalloproteinases as Targets for Anti-Inflammatory Drugs,
 K.M.K. Bottomley, D. Bradshaw, J.S. Nixon (Editors), 1999
Free Radicals and Inflammation, P.G. Winyard, D.R. Blake, C.H. Evans (Editors), 1999
Gene Therapy in Inflammatory Diseases, C.H. Evans, P. Robbins (Editors), 2000
New Cytokines as Potential Drugs, S. K. Narula, R. Coffmann (Editors), 2000
High Throughput Screening for Novel Anti-inflammatories, M. Kahn (Editor), 2000
Immunology and Drug Therapy of Atopic Skin Diseases,
 C.A.F. Bruijnzeel-Komen, E.F. Knol (Editors), 2000
Novel Cytokine Inhibitors, G.A. Higgs, B. Henderson (Editors), 2000
Inflammatory Processes. Molecular Mechanisms and Therapeutic Opportunities,
 L.G. Letts, D.W. Morgan (Editors), 2000

Cellular Mechanisms in Airways Inflammation, C. Page, K. Banner, D. Spina (Editors), 2000
Inflammatory and Infectious Basis of Atherosclerosis, J.L. Mehta (Editor), 2001
Muscarinic Receptors in Airways Diseases, J. Zaagsma, H. Meurs, A.F. Roffel (Editors), 2001
TGF-β and Related Cytokines in Inflammation, S.N. Breit, S. Wahl (Editors), 2001
Nitric Oxide and Inflammation, D. Salvemini, T.R. Billiar, Y. Vodovotz (Editors), 2001
Neuroinflammatory Mechanisms in Alzheimer's Disease. Basic and Clinical Research, J. Rogers (Editor), 2001
Disease-modifying Therapy in Vasculitides, C.G.M. Kallenberg, J.W. Cohen Tervaert (Editors), 2001
Inflammation and Stroke, G.Z. Feuerstein (Editor), 2001
NMDA Antagonists as Potential Analgesic Drugs, D.J.S. Sirinathsinghji, R.G. Hill (Editors), 2002
Migraine: A Neuroinflammatory Disease? E.L.H. Spierings, M. Sanchez del Rio (Editors), 2002
Mechanisms and Mediators of Neuropathic pain, A.B. Malmberg, S.R. Chaplan (Editors), 2002
Bone Morphogenetic Proteins. From Laboratory to Clinical Practice, S. Vukicevic, K.T. Sampath (Editors), 2002
The Hereditary Basis of Allergic Diseases, J. Holloway, S. Holgate (Editors), 2002
Inflammation and Cardiac Diseases, G.Z. Feuerstein, P. Libby, D.L. Mann (Editors), 2003
Mind over Matter – Regulation of Peripheral Inflammation by the CNS, M. Schäfer, C. Stein (Editors), 2003
Heat Shock Proteins and Inflammation, W. van Eden (Editor), 2003
Pharmacotherapy of Gastrointestinal Inflammation, A. Guglietta (Editor), 2004
Arachidonate Remodeling and Inflammation, A.N. Fonteh, R.L. Wykle (Editors), 2004
Recent Advances in Pathophysiology of COPD, P.J. Barnes, T.T. Hansel (Editors), 2004
Cytokines and Joint Injury, W.B. van den Berg, P. Miossec (Editors), 2004
Cancer and Inflammation, D.W. Morgan, U. Forssmann, M.T. Nakada (Editors), 2004
Bone Morphogenetic Proteins: Bone Regeneration and Beyond, S. Vukicevic, K.T. Sampath (Editors), 2004
Antibiotics as Anti-Inflammatory and Immunomodulatory Agents, B.K. Rubin, J. Tamaoki (Editors), 2005
Antirheumatic Therapy: Actions and Outcomes, R.O. Day, D.E. Furst, P.L.C.M. van Riel, B. Bresnihan (Editors), 2005
Regulatory T-Cells in Inflammation, L. Taams, A.N. Akbar, M.H.M Wauben (Editors), 2005
Sodium Channels, Pain, and Analgesia, K. Coward, M. Baker (Editors), 2005
Turning up the Heat on Pain: TRPV1 Receptors in Pain and Inflammation, A.B Malmberg, K.R. Bley (Editors), 2005
The NPY Family of Peptides in Immune Disorders, Inflammation, Angiogenesis and Cancer, Zofia Zukowska, Giora Z. Feuerstein (Editors), 2005
Complement and Kidney Disease, P.F. Zipfel (Editor), 2006